Springer Tracts in Advanced Robotics 101

For further volumes:
http://www.springer.com/series/5208

STAR (Springer Tracts in Advanced Robotics) has been promoted under the auspices of EURON (European Robotics Research Network)

Foreword

Robotics is undergoing a major transformation in scope and dimension. From a largely dominant industrial focus, robotics is rapidly expanding into human environments and vigorously engaged in its new challenges. Interacting with, assisting, serving, and exploring with humans, the emerging robots will increasingly touch people and their lives.

Beyond its impact on physical robots, the body of knowledge robotics has produced is revealing a much wider range of applications reaching across diverse research areas and scientific disciplines, such as: biomechanics, haptics, neurosciences, virtual simulation, animation, surgery, and sensor networks among others. In return, the challenges of the new emerging areas are proving an abundant source of stimulation and insights for the field of robotics. It is indeed at the intersection of disciplines that the most striking advances happen.

The *Springer Tracts in Advanced Robotics (STAR)* is devoted to bringing to the research community the latest advances in the robotics field on the basis of their significance and quality. Through a wide and timely dissemination of critical research developments in robotics, our objective with this series is to promote more exchanges and collaborations among the researchers in the community and contribute to further advancements in this rapidly growing field.

The book *Introduction to Humanoid Robotics* by Shuuji Kajita, Hirohisa Hirukawa, Kensuke Harada and Kazuhito Yokoi is an enriched English translation of a former book in Japanese. The six-chapter collection offers a complete treat on the fundamental methodologies and technologies of humanoid robotics; namely, kinematics, dynamics, biped walking, motion generation and simulation. Even though the contents are focused on the achievements of a huge research effort on the Humanoid Robotics Project undertaken at the Advanced Institute of Science and Technology in Tsukuba, the material is of wide interest for virtually any scholar wishing to pursue work in this fascinating field.

The first contribution to the series on humanoid robots, this volume constitutes a very fine addition to STAR!

Naples, Italy
November 2013

Bruno Siciliano
STAR Editor

Foreword

Preface to English Edition

This book is based on our Japanese text book simply titled "Humanoid robots." We wrote the book because we wanted to compile basic knowledge of analysis and control of humanoid robots in a compact form just for our small group. It was our surprise and pleasure to learn that the book was also useful for other people, mostly those who wanted to start research in the field of humanoid robotics. Translations in Chinese, German and French have already been published. For the Chinese translation, we express our appreciation to Professor Yisheng Guan at Guangdong University of Technology. For the French translation, we wish to show our appreciation to Professor Sophie Sakka at Ecole Centrale de Nantes. Finally, we can release the English version with great help from Mr. Hajime Saito from the Kawada Robotics Corporation and Professor Bill Goodwine at the University of Notre Dame.

Although the original book was written nine years ago, we believe the basic outline of this book is still useful. To include recent results from our group and others, we have added new sections to Chapters 1, 2 and 4.

A realistic humanoid robot is intriguing to many people no matter their culture, for it has long been a dream of all people. The challenge to create a life-mimicking machine had started thousands of years ago. Although, we see impressive development now, I never think scientists are close to the goal. All robotics researchers must sigh over their incompetence looking at the exquisite mechanism, ourselves. Therefore, we are in the middle of a long journey. May the journey leads us a prosperous, pleasant and exciting future!

September 2013 On behalf of authors, Shuuji Kajita

Preface to the Original Japanese Edition

Recent humanoid robots which appear in televisions and exhibitions can walk and perform impressive dances as if they have actors inside. Many people might say, "Wow, its wonderful! But how can they do that?" The primary goal of this book is to answer such questions. The theories and technologies introduced in this book are actually used to control our humanoid robot, HRP-2. Similar technologies are also used for other famous humanoid robots like Honda's ASIMO and Sony's QRIO.

If you quickly browse through this text, you will find this book is not easy to read. Indeed, it is full of equations and other math, which may prove to be a lethal dose to those with allergic reactions to mathematics. To soften such an insipid impression and to aid the reader in interpreting and understanding the equations we put pictures and drawings whenever possible. However we want to emphasis that those are like *music scores* composed by the words of mathematics and physics, which are indispensable to enable such impressive humanoid robot technology. We also expect that many readers interested in humanoid robots may come to recognize the importance and indispensable role of science and technology which truly supports our modern society.

The first chapter was written by Hirohisa Hirukawa, the leader of the Humanoid Robotics research Group (HRG) in the Intelligent Systems Research Institute (ISRI) of AIST (National Institute of Advanced Industrial Science and Technology). Chapter 3 is written by Kensuke Harada and Shuuji Kajita in HRG. Chapter 5 was written by Kazuhito Yokoi, the leader of the Autonomous Behavior Control Research Group in ISRI. Chapter 2, 4 and 6 were written by Shuuji Kajita.

We could not publish this book without help of many other people. First of all, we express our thanks to Tadahiro Kawada, Takakatsu Isozumi and other excellent engineers of Kawada Industries, Inc. who have designed and built marvelous hardware, including the humanoid robot HRP-2. We also would like to thank to Token Okano and Yuichiro Kawasumi of General Robotix, Inc. (GRX) who help us everyday by maintaining our robots. We thank to Kenji Kaneko, Fumio Kanehiro, Kiyoshi Fujiwara, Hajime Saito and

Mitsuharu Morisawa in HRG, who have been producing excellent research results. The contents of this book depend on the results of the all HRG members. Moreover, they gave much valuable advice for the draft of this book. Haruhisa Kurokawa, the leader of Distributed System Design research group in ISRI of AIST gave us much important advice for our final draft. Takashi Nagasaki, who was an undergraduate student of Tsukuba University pointed out many errata. If there were something valuable in this book, the authors owe it to the above mentioned people. Of course, it is needless to say, we are fully responsible for all errata which might still exist in this book.

Finally, we thank to Shigeoki Hirai, the research director of ISRI and Kazuo Tanie, the former research director of ISRI. We could not complete this book without having their generous management guidance and support.

December 2004 On behalf of authors, Shuuji Kajita

Contents

Chapter 1
Introduction

A humanoid robot is a robot that has a human-like shape. Since many robots in scientific fictions look like humans, a humanoid robot may be the default of robots for most people. On the other hand, it is difficult to claim that robots should be humanoid robots which are supposed to do some tasks in the real world, considering that aircrafts do not look like birds. The required functions for a robot may determine the optimal shape of the robot.

We have to consider what we expect from robots before we investigate what is the optimal shape of robots. An automobile had become the product that created the largest industry in the 20th century, since it satisfied human desires to go to far places and to enjoy driving itself. We should consider what kinds of the desires robots can satisfy. We claim that robots should be expected to do tasks which we do not want to do and to be our partners to enjoy communications. Considering how to realize the functions of robots, the features of humanoid robots can be summarized as follows; 1. humanoid robots can work in the environment for humans as it is, 2. humanoid robots can use tools for humans as it is, and 3. humanoid robots has a human-like shape.

Let us examine the first feature. The environment of the modern society is designed for humans. For example, the width of corridor, the height of a stair, and the position of a handrail are determined to fit the size and motions of humans. Therefore, we need not modify the human environment for a robot to operate when the robot has a human shape and move like a human. An uneven floor has to be made flat, a narrow passage should be removed and a lift must be available when a robot moves on wheels. It should be more economical to develop humanoid robots than to modify the whole environment.

The second feature should imply a similar effect. Most of tools for humans are designed to be used by humans. For example, the size and shape of chairs are determined to sit on them, and the height of dining tables are decided to eat on them. A driver's cockpit is designed to control a car. The shape of a screw driver or scissors can be operated best by articulated fingers. The tools

S. Kajita et al., *Introduction to Humanoid Robotics*,
Springer Tracts in Advanced Robotics 101,
DOI: 10.1007/978-3-642-54536-8_1,

for humans are likely to be used by humanoid robots as it is. It should be more economical to used humanoid robots than to re-design numerous tools.

A similar discussion is written in a novel 'The Caves of Steel' by Issac Asimov. A world famous professor explains why robots should be humanoid robots in the novel and has the similar conclusions with us. Honestly speaking, we need years to reach the conclusions. It is amazing that the same conclusions were already obtained in a fifty-years old novel.

The third feature must need some explanation. A robot is easily personified when the robot looks like a human. The further a robot is from a human shape, the less humans feel a human in the robot. It is fun to watch a biped humanoid robot is dancing, but the dancing of a wheeled robot should be less attractive. A human-like shape is very important to realize a partner robot that can make us enjoy. The third feature must be the primary account why many robots look like humans in scientific fictions. In the real world, WABOT-1 was developed by Ichiro Kato et al. from Waseda University in 1973 (see Fig.1.1).

WABOT-1 could recognize objects by vision, understand spoken language, speak by artificial voice, manipulate the objects by two hands, and walk on biped legs, though the level of the technologies was not so matured. Therefore, it is reasonable to call WABOT-1 *the first humanoid robot.* Ichiro Kato's group also developed WABOT-2 in 1984 which was able to play a piano (see Fig.1.1)[68]D WABOT-2 had played a piano at Tsukuba Science Expo'85 in Japan.

The epoch of humanoids was opened by the astonishing reveal of Honda humanoid P2 in 1996. Honda started a confidential project of humanoid robots in 1986 when one year had passed after WABOT-2 played a piano. P2, 180 cm height and 210 kg weight, is the first humanoid robot that can walk on

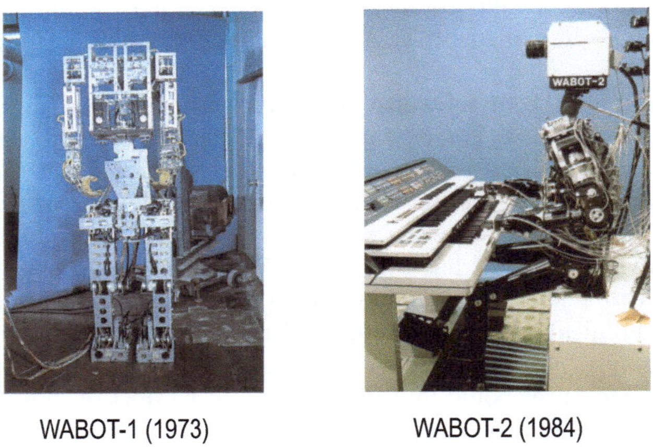

Fig. 1.1 Humanoid robots from Waseda University
(Coutesy of Humanoid Robotics Institute, Waseda University)

biped legs with a sufficient stability and mount a computer and battery on the body. Honda published P3, 160 cm height and 130 kg weight, in 1997, and ASIMO, 120 cm weight and 43 kg weight, in 2000. The pictures of P2, P3 and ASIMO are shown in Fig.1.2.

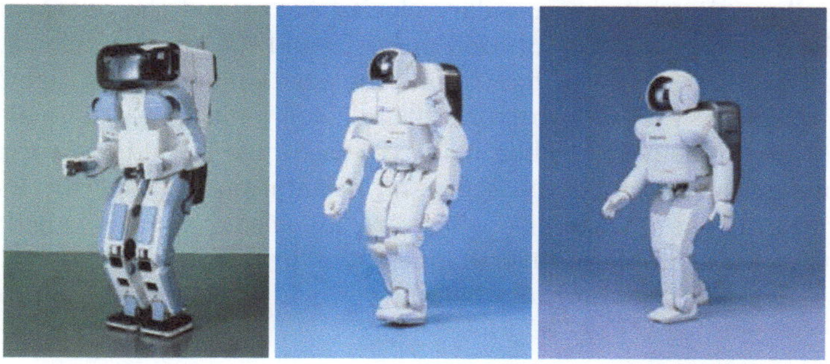

P2 (1996) P3 (1997) ASIMO (2000)

Fig. 1.2 Humanoid robots from Honda (Coutesy of Honda)

Before P2 was published, the majority in robotics community were pessimistic about the development of a biped humanoid robot that can walk stably. This is the reason why Honda P2 astonished the community. Then what are the major differences between the conventional humanoid robots and P2? Let us examine the hardware at first.

Most humanoid robots developed at universities were made by graduate students or by a small manufacturer. Then the mechanical links of the robots had to be made by bending or cutting and the whole structure was not rigid enough. The reduction mechanisms were implemented by heavy gears with large backlash. By contrast with the old robots, Honda humanoid robots use casted mechanical links with high rigidity and light weight using most advanced mechanical CAD. It was obvious that the casted links should have such properties, but the links were too expensive to be developed by university projects. Honda humanoid robots use harmonic drives which have no backlash. The conventional harmonic drives could transform too little torque to be applied to biped walking, so Honda developed harmonic drives with high torque capacity. After Honda P2 was revealed, most advanced humanoid robots have comparable configurations with those of Honda humanoid robots.

Let us consider the sensors for the robots. Biped walking may not be stable due to disturbance even when the desired walking pattern is planned to make the walking stable. Then the walking should be stabilized by a feedback control that demands appropriate sensors. The humanoid robots developed in the early stage did not have necessary sensors, but Honda humanoid robots have accelerometers and gyroscopes, to find the orientation of bodies and six-axes force/torque sensors to find the contact force/torque between the feet and the floor.

The goal of this textbook is to give the theoretical background for the development of the software to control the well-designed hardware described above.

Chapter 2 overviews the kinematics of humanoid robots. A representation of the motions of the robots is presented after the representation of rotations in three dimensional space, angular velocity vector and the relationship between the derivatives of the rotation matrices and the angular velocity vector are described. It is presented how to find the position and orientation of a link such as a hand or a foot of the robot from given joint angles. The method is called forward kinematics. Then it is explained how to find the corresponding joint angles from given position and orientation of a specific link. The computation is the inverse of the forward kinematics, and is called inverse kinematics. An example of inverse kinematics problems is illustrated in 1.3. When the configuration of the robot shown in 1.3(a) is given, the problem is

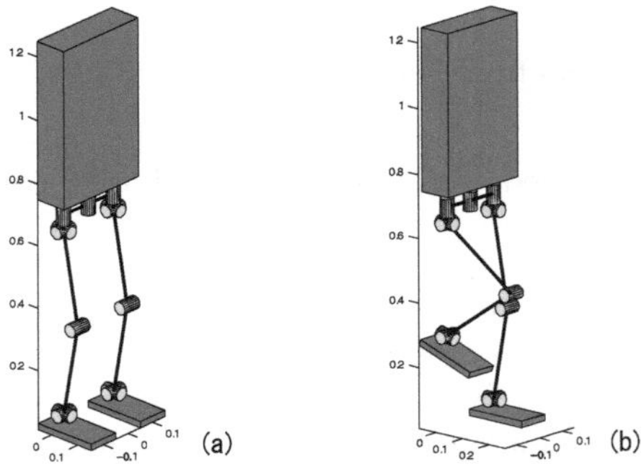

Fig. 1.3 Example of inverse kinematics of a biped robot; (a) initial configuration, (b) that at which the right foot is raised by 0.3 [m] and rotated by 20 [deg] about y-axis

how to find the corresponding joint angles at which the right foot is raised by 0.3 [m] and rotated by 20 [deg] about Y axis shown in 1.3(b).

Generally speaking, the postion and orientation of a link and the joint angles are represented by nonlinear equations, since most joints of robots are rotational ones. The inverse kinematics problem can be solve by finding solutions of the nonlinear equations analytically, but it is unlikely to solve the nonlinear equations with many variables and a high Bezout number even if the rotation is parametarized algebraically. However, the relationship between the derivatives of the postion and rotation of a link and those of the joint angles can be represented by linear equations, and the inverse kinematics problem can be solved by finding solutions of the linear equations and integrating the solutions. The coefficient matrix of the linear equations is called Jacobian, which is an important concept in many fields including robotics.

Chaper 3 explains the concept of ZMP (Zero-Moment Point) that plays an important role in the motion control of humanoid robots. When the robot is falling down, the sole of the supporting foot should not contact with the ground any more. The ZMP (Zero Moment Point) proposed by Vukobratović et al. is a criterion to judge if the contact between the sole and the ground can be kept without solving the corresponding equations of motions. The contact is kept if the ZMP is an internal point on the sole. When the robot does not move, the contact is kept when the projection of the center of the mass of the robot onto the ground is an internal point of the sole. See Fig.1.4. The

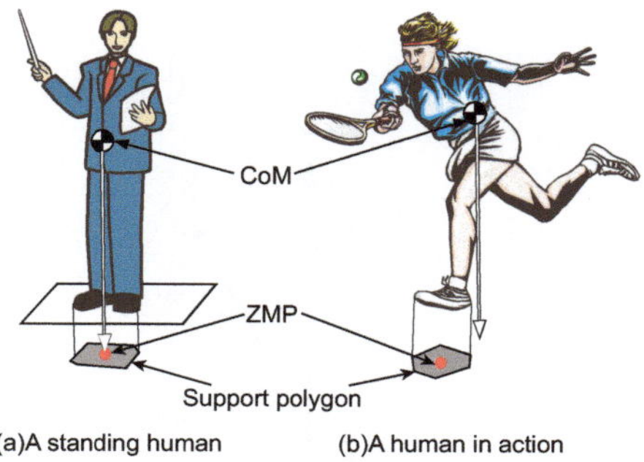

Fig. 1.4 Center of the mass, ZMP and supporting polygon

ZMP can be considered to be a dynamic extension of the projection.

The ZMP can be used to plan motion patterns that can make the robot walk while keeping the contact between the sole of the supporting foot and the ground. The walking patterns of the majority of humanoid robots have been generated based on the ZMP after Honda P2 appeared.

The robot may not fall down even when the sole of the supporting foot left the ground. It can keep walking or standing by controlling the swing leg and changing the touch down positions. The ZMP criterion is a sufficient condition to prevent the robot from falling down, and not a necessary one. It can neither judge rigorously if the contact is kept when the robot walks on a rough terrain or a stair. When the robot walks while catching a handrail, the contact may be more stable but the ZMP criterion can not tell how much the contact should be made stable. Several trials have been made to extend the criterion, but the generic and rigorous criterion has not been established so far. Chapter 3 presents the concept of the ZMP, the relationship between the contact force and the ZMP, the sensing of the ZMP and an algorithm to compute the ZMP based on the forward dynamics of humanoid robots, which is overviewed as well.

Chapter 4 describes how the walking patterns of biped robots can be generated and the walking can be controlled. Generally, the walking patterns are planned to make the robots walk with no disturbance at first, and a feedback control is applied to stabilize the motions. Various methods have been proposed to general the walking patterns. One method is based on the dynamics of the linear inverted pendulum whose height of the center of the mass is kept by controlling the contact force and the length of the pendulum. Another one generates the patterns using the ZMP as the criterion to judge the contact stability.

Chapter 4 starts from the introduction of a method based on two dimensional linear inverted pendulum, the method is extended to that based on three dimensional one, and it is applied to generate the patterns of multi-link models. Fig.1.5 shows the concept of three dimensional linear inverted pendulum.

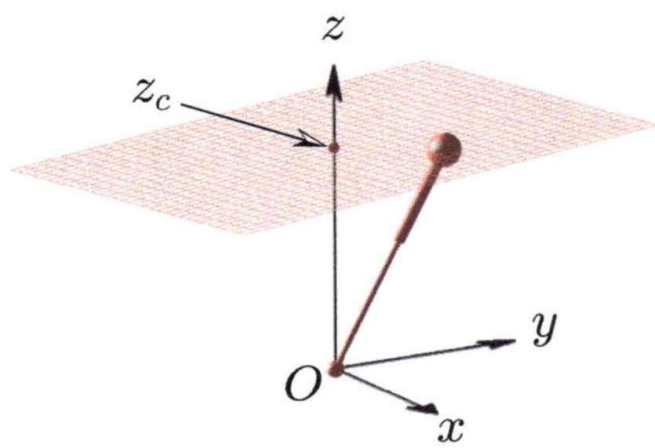

Fig. 1.5 Concept of three dimensional linear inverted pendulum - the motions of the center of mass is constrained on a specified plane by controlling the contact force and the orientation of the plane is independent to the motions of the center of mass

The ZMP-based method is overviewd as well. The relationship between the derivatives of the joint angles and the ZMP can be given by nonlinear differential equations. It is called the ZMP equations. It is difficult to find trajectories of the joint angles which let the ZMP follow specified ZMP ones due to the nonlinearity. The ZMP equations were simplified under several assumptions and their solutions were found by a batch processing in the early stage. The walking patterns were computed in an offline fashion, even when Honda P2 was published. Honda developed a realtime method to find the patterns and applied it to ASIMO. Nishiwaki et al. invented another realtime algorithm to solve the equations by constraining the motions of the waist joint onto a horizontal plane. Kajita et al. solved the linearized equations in realtime by a preview control and applied it to humanoid robot HRP-2. Chapter 4 explains the methods.

Even when the motion patterns are carefully planned to make the walking stable, humanoid robots may still tip over due to disturbances caused by terrains of the floor, low rigidness of the mechanical structure and the backlash of reduction gears. Therefore, it is demanded to find the status of the robot by sensors including the orientation of the body by an accelerometer and a gyroscope and the contact force and torque of the feet by a force/torque sensor and to apply some feedback control to stabilize the motions. The present configurations of the feedback controllers are the combinations of the orientation control of the body, the control of the center of mass, the compliance control of the feet contact, the impact force control of the foot touch down and so on. The fine tuning of the feedback controllers has become possible since the significant progresses were made in the hardware of the robots as mentioned above. The chapter overviews the principle of the feedback controllers. Fig.1.6 shows the motions of the feet of humanoid robot HRP-2 which walks on a rough terrain with the maximum height of the ramps 2 [cm] and the maximum inclination of the slopes 5 %. The feedback controls are definetely necessary to realize the walking.

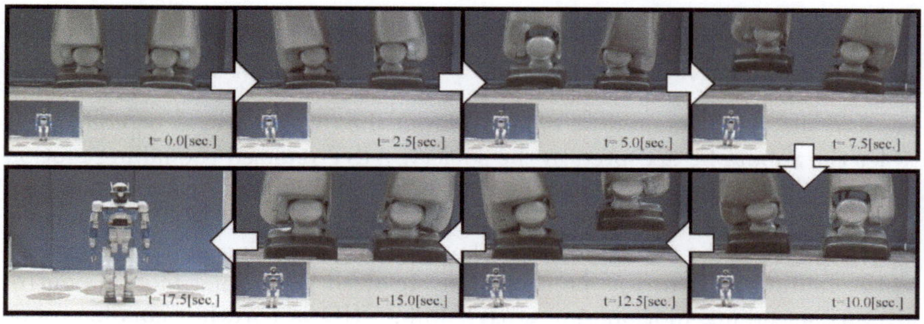

Fig. 1.6 Feet motions of HRP-2 walking on a rough terrain

Chapter 5 presents how the whole body motions, other than biped walking, of humanoid robots can be realized. Humanoid robots may lie down, get up, carry an object, go through a narrow place and dance. AIST realized the first human-sized humanoid robot that can lie down and get up, which is shown in Fig.1.7.

This chapter overviews how various whole body motions can be generated and controlled. It is described how gross motions of the robots can be generated, including the methods using motion capture systems, graphics user interface and the searching of the configuration spaces of the robots. The methods generate the motions with considering neither the dynamics of the robots nor the contact stability between the robots and the environments, and therefore the motions may not be executed by real robots and the robots may fall down in most cases. Besides, the configurations of the humans whose motions were captured should not identical with those of the real robots. Various methods have been proposed to solve the problems, including the dynamics filters and feedback controllers. The chapter covers the teleoperations of humanoid robots as well.

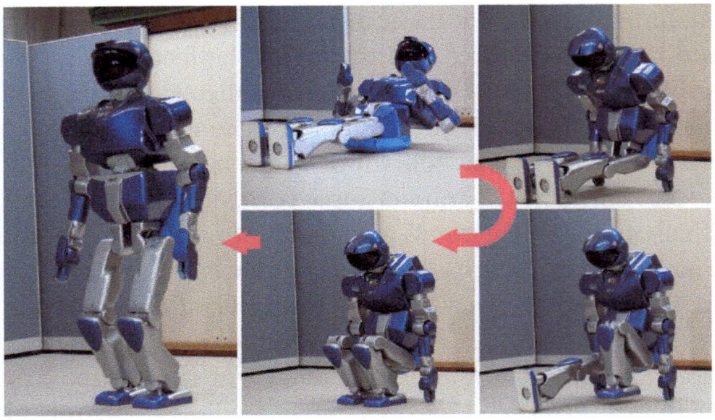

Fig. 1.7 Getting up of humanoid robot HRP-2P

Though biped robots can go up and down stairs, go over ramps and go through narrow places, the robots may fall down due to the relatively high center of the mass and smaller footprints with serious damage. The advantages of the robots should be enhanced and the disadvantages be conquered to let the robots be accepted in the society. AIST realized the falling motion control of human-sized humanoid robots in Feb. 2003. The motion is like Ukemi motions of Judo that can minimize the damage of the body when a player is thrown by the opponent. The controllable falling motion is limited to the falling backward at present. Sony realized the falling motion control for various falling motions for QRIO, but QRIO is a rather smaller robot than the human-size. The impact of the touch down is significantly larger when a human-sized humanoid robot is falling down, and we should have a tough hardware and a better controller to handle it. Chapter 5 describes the falling motion controller as well as the method to realize the lying down and the getting up.

Chapter 6 presents the algorithms for the dynamics simulation of humanoid robots. The forward dynamics of a robot is the problem to find the updated state of the joints of the robot when the current state of the robot and the generalized force to the joints are given. The inverse dynamics of a robot is the problem to find the generalize force to the joints to realize the desired updated state of the robot. The chapter focuses the attention to the forward dynamics, which starts from that of a rigid body rotating in a gravity-free space and extends the formulation to include the translational motions of the body. An example of the computed motion is shown in Fig.1.8. Finally, we consider the forward dynamics of conneced bodies i.e. robots, which enables the readers to know how the dynamic simulation of humanoid robots can be implemented. The method is based on the Newton-Euler equations, and efficient algorithms to solve the equations were proposed in 1980s. An algorithm developed by Fetherstone will be described as a representative one.

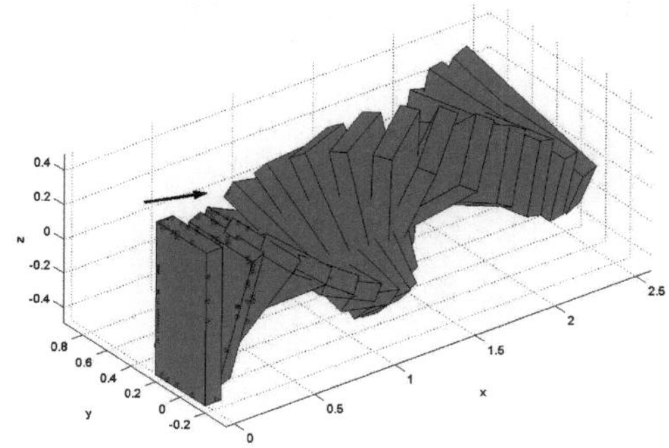

Fig. 1.8 A computed motion of a rigid body in a gravity-free space

It was the summary of the textbook. The objective of the textbook is to give the foundation to develop the software for controlling the various motions of humanoid robots.

In the following, a perspective of the future of humanoid robotics is overviewed. A humanoid robot can be an integration platform for various robotic technologies, since it has two arms and two legs as well as visual and audio sensors. However, the available computing resource and sensors on the robot should be rather limited due to the space constraint of the robot. We have to make the computers and sensors more compact and powerful to fix the problem. Then more intelligence can be integrated on the platform, and the focus of humanoid robotics may shift from the mobility to the intelligence and the applications based on it.

METI (Ministry of Economy, Trade and Industries) of Japan had run Humanoid Robotics Project (HRP for short) from 1998FY to 2002FY. The leader of HRP was Hirochika Inoue from the University of Tokyo, and HRP developed the fundamental technologies for humanoid robots and explored the applications.

The first feature of humanoid robots is "humanoid robots can work in the environment for humans as it is", and the maintenance tasks of an industrial plant had been investigated as an example application of the feature. Humanoid robot HRP-1 was able to execute the tasks in a mockup of an industrial plant which includes staris, ramps and pits. Fig.1.9 shows HRP-1 going down stairs in the mockup.

The second feature is "humanoid robots can use tools for humans as it is", and the driving of an industrial vehicle was examined as an example of its applications. The idea is to realize a teleoperation system for an industrial vehicle by making a humanoid robot drive an industrial vehicle and by controlling the robot by a human operator in a remote space, which can be applied to rescue works. Fig.1.10 shows that a backhoe is drived by humanoid robot HRP-1S which is teleoperated by a human operator. HRP-1S wears a

Fig. 1.9 HRP-1 going down stairs

Fig. 1.10 HRP-1S driving a backhoe

waterproof suit. Some amounts of waterdrop can be seen in the picture by a carefull observation.

The third feature is "humanoid robots has a human-like shape", and a trivial application of the feature is entertainment. Honda ASIMO and Sony Qrio have appeared in many commercial messages, and a traditional Japanese dance is reproduced by HRP-2 to realize a digital archive of the dance culture. Another possible application of the feature is a human simulator to evaluate tools for humand like a driver cockpit of vehicles and welfare apparatuses.

Humanoid robotics is in the first exiting epoch after 1996, but it is still very difficult to realize significant products using humanoid robots without passing through a nightmare period.[1] Since humanoid robotics demands a large scale investiment, we have to realize its new applications every five years to continue the efforts. If we aim at the realization of a humand robot whose

[1] Most innovations had to pass through nightmare periods between that of the fundamental research and that of the industrialization, in which the new technologies were criticized including CAD and industrial robots. Hiroyuki Yoshikawa named the period "a nightmare period" which is an important stage to grow innovations.

ability is comparable with that of a human, we may need a century. We should need a decade at least even if we go ahead to realize an autonomous humanoid robot. From the viewpoint, we would like to propose the goal of 2010 as follows; a humanoid robot that can walk on floors in the daily environment, go up and down stairs and ladders, plan its paths autonomously, fall down without a serious damage, get up from the floor, step over small obstacles, pass through narrow spaces, open/close doors, and manipulate an object by one hand while supporting its body by another hand. It is a humanoid robot that can go any place where a normal human can go. When the mobility is realized, the possible applications of the robot should include the maintenance tasks of industrial plants and the management of hazadous objects. Among the goals, it is most difficult for a humanoid robot to fall down without a serious damage. Even when an appopriate control is applied to the robot when it falls down, the current hardware of the robot is too fragile to keep the mobility. We need much more works to conquer the problem.

Next, as the goal of 2015, we propose the development of autonomous humanoid robots which are able to execute rather simple tasks autonomously that can be done easily by humans. To this end, humanoid robots must have theree dimensional vision that can know the shape, postion and orientation of an object, a dexterous hand that can manipulate various kinds of objects, force/torque sensors that can know the state of the manipulated objects, motion planning and so on. Then the possible applications include assembly of mechanical structures and unregular manipulation tasks. It may be possible to produce more than one thousand copies of the robot when the applications can be realized, and we can claim that we already passed through "the nightmare period" of humanoid robotics.

As the goal of 2020, we propose the development of a humanoid robot that can work cooperatively with humans while sharing the common space with them. The final goal of HRP can be attained when the goal of 2020 is reached. To this end, humanoid robots must have the intelligence for safety as well as high autonomy. It is very difficult to realize the safety, since rather large power should be required for the manipulation and the mobility. It can be understood more comprehensively when we remember that even humans can hart others when a narrow space is shared. The robot should be more safe than a human to be accepted by the society. The efforts like the minimization of the power or the coverage by a soft material are not enough for the purpose, and more sophisticated technologies should be integrated like the realtime observation of the environemnt and the safety control of arms and legs. When the safety intelligence is integrated, it may give a chance to realize the applications like the human care services examined in HRP. It is most difficult to be realized, but the expected sized of the corresponding market should be largest, since the robot can be used at home then. "A humanoid robot for every home" may not be a dream when the mission was completed.

It is very difficult to reach the goal of 2010 from the state of the art in 2005. There is no clear roadmap to attain the goal of 2015. The goal of 2020 is just a dream at present. The objective of this textbook is to let more people learn the foundations of humanoid robotics and spread the use of humanoid robotics. It should be fantastic news for scientists and engineeers that most work in the field of humanoid robotics is still waiting to be done. The development of the automobile replaced the horse as a mobility tool, and the development of humanoid robots offers the promise to replace humans as the bearer of hard and dull tasks. Clearly the goal of humanoid robots alleviating the more burdensome tasks currently performed by humans is both more difficult and potentially more significant than the example of the automobile. We wish that this text may contribute to the ability and inspiration of our colleagues to strive for such a lofty goal.

The Development at AIST Since 2005

In this section, we present the development at AIST (National Institute of Advanced Industrial Science and Technology) since the previous section was written.

In 2005, we developed biped dinosaur robots with the support of New Energy and Industrial Technology Development Organization (NEDO) and AIST. The purpose was for exhibitions at the 2005 World Exposition Aichi. At the same time, our intention was to seek possible applications of biped technologies in the entertainment industry.

Figure 1.11 shows the developed dinosaur robots [38]. The Tyrannosaurus Rex robot has a body length of 3.5 m from the head to the tail, a weight 83 kg, and 27 DoF (degrees of freedom). The Parasaurolophus robot has the same body length, but it has a weight of 81 kg, and 26 DoF. They are 1/3.5 scale of the real dinosaurs. During the exposition of half a year, these robots successfully performed 1,812 demonstrations and entertained the audience.

As explained in the former section, the Ministry of Economy, Trade and Industry of Japan conducted the Humanoid Robotics Project (HRP) from FY (fiscal year) 1998 to FY 2002. One of the project's outcomes was the humanoid robot HRP-2, which was developed by Kawada Industries Inc., Yaskawa Electric Corporation, the Shimizu Corporation, and AIST [65]. Throughout this book, we will use the HRP-2 as a typical robot to explain the basics of humanoid robotics.

On the other hand, a HRP-2 robot has limitations in its manipulation ability and practical working environment, such as a construction site. To address these limitations, a new humanoid robot HRP-3 was developed in 2007 by Kawada Industries Inc., Kawasaki Heavy Industries, and AIST. This project was supported by NEDO [67]. Figure 1.12 shows the HRP-3, which is a humanoid robot of 1606 mm height, 68 kg weight, and 42 total DoF. For better manipulation ability, the robot has 7 DoF arms and 6 DoF hands (the HRP-2 has 6 DoF arms and 1 DoF hands). In addition, the whole robot and hardware was designed to be dust proof and splash proof in consideration of various environments.

(a) Walking Tyrannosaurus rex robot (Courtesy of AIST)

(b) Parasaurolophus robot (c) The robot and a human

Fig. 1.11 The dinosaur robots

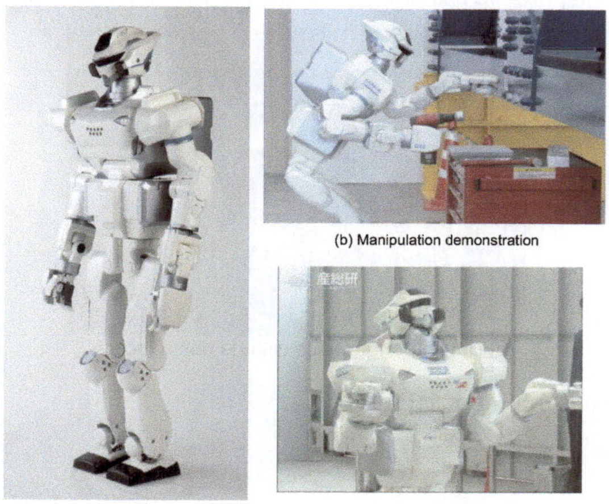

(a) HRP-3 (b) Manipulation demonstration (c) Splash-proof demonstration

Fig. 1.12 HRP-3 and its demonstrations

In 2009, we developed a new humanoid robot, Cybernetic human HRP-4C. This robot was designed to have body dimensions close to average Japanese young female [63]. In Fig.1.13(a), we can see HRP-4C has a much more human-like appearance (left) than our previous HRP-2 (right). The purpose of this development was to seek humanoid applications in the entertainment industry, for example, fashion shows. We also intended the use the robot to evaluate devices for humans. Using HRP-4C, we realized human-like walking with toe supporting period (Fig.1.13(b)) [69]. Figure 1.13(c) shows the performance where HRP-4C dances with human dancers. For such performances, we developed a new software for efficient choreography [122]. The hardware of HRP-4C has been modified since 2009, and its current specification is 160 cm height, 46 kg weight and total 44 DoF [64].

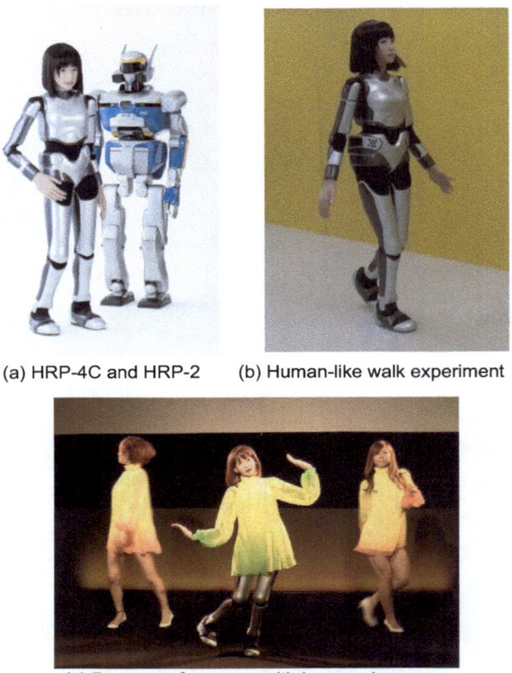

(a) HRP-4C and HRP-2 (b) Human-like walk experiment

(c) Dance performance with human dancers

Fig. 1.13 Cybernetic human HRP-4C

In 2010, another humanoid robot HRP-4 was developed by Kawada Industries, Inc. and AIST. Figure 1.14 shows the robot which has a 151 cm height, 39 kg weight and 34 DoF [64]. HRP-4 was designed as a R&D platform which has a lightweight and slim body compared with our former platform HRP-2. To realize object manipulation better than HRP-2, the robot has 7 DoF arms and 2 DoF hands.

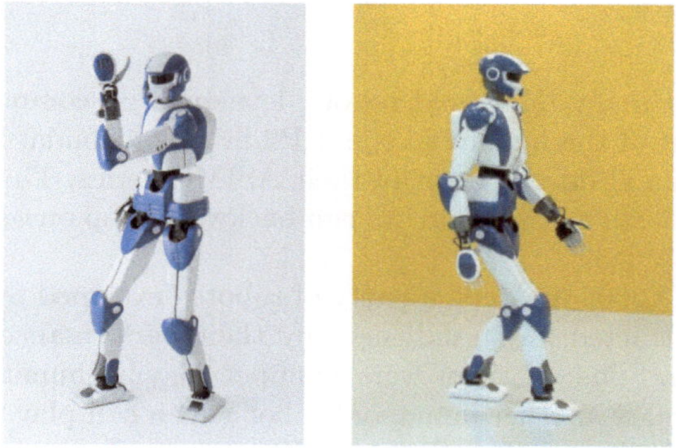

(a) HRP-4 (b) HRP-4 walk experiment

Fig. 1.14 HRP-4 and its walk experiment

Universities and Research Institutes

Research activities on biped humanoid robots around the world have accelerated in the last decades. In Japan, professor Takanishi's group in Waseda university has been actively developing many biped humanoid robots following professor Kato who built world's first humanoid robot WABOT-1. In 2006, their WABIAN-2R demonstrated impressive human-like walking with knee stretched, heel-contact and toe-off motion [142].

Another prominent group is led by professor Inaba in University of Tokyo. In 2010, they demonstrated HRP-2 which can handle objects of unknown weight based on online estimation of the operational force [123]. They are also developing their original biped robot which can balance even when kicked or otherwise disturbed [52].

ATR Computational Neuroscience Laboratories is studying humanoid robots from a viewpoint of brain science. Using the humanoid robot CB-i developed by SARCOS Inc., their biologically feasible balance controller has been tested [95].

Needless to say, biped humanoid research is not limited in Japan. As remarkable examples, we can see LOLA by Technische Universität of München (TUM) [119], HUBO2 by Korea Advanced Institute of Science and Technology (KAIST) [12], BHR-2 by Beijing Institute of Technology [148], iCub by Italian Institute of Technology (IIT), the University of Genoa [34], CHARLI by Virginia Polytechnic Institute and State University [22], and TORO by the German Aerospace Center (DLR) [15, 16].

Companies

There also exist many humanoid robots developed by companies. Since the surprising debut of the humanoid robot P2 in 1996, Honda, has been carrying on research and development of their ASIMO series. The latest ASIMO unveiled in 2011 can run at 9 km/m, run backward, hop on one leg or on two legs continuously [43].

At the EXPO 2005 in Aichi, a party of robots developed by Toyota Motor Corporation attracted large audiences by their performance in the Toyota Group Pavilion. Some of them were trumpet playing humanoid robots. In 2007, they revealed another humanoid robot which can play the violin [17].

A South Korean company Samsung Electronics, has been also developing humanoid robots with the Korean Institute of Science and Technology (KIST). Their latest humanoid robot is Roboray, which can perform knee-stretched human-like walk [13].

In 2012, an American robotics design and engineering company, Boston Dynamics, developed a humanoid robot PETMAN to test chemical protective clothing [35]. Powered by hydraulic actuators and controlled by advanced control software, this robot can perform squats, squats while turning and side-steps with its arms raised overhead, as well as natural human-like walking of up to 4.8 km/hr.

We cannot purchase above mentioned robots for they were developed as a part of the long range R&D projects. On the other hand, there already exist commercially available humanoid robots for research purposes. For example, Kawada Industries is selling the humanoid robot HRP-4 as a research platform [56]. PAL Robotics in Barcelona has also developed a humanoid robot REEM-C for sale [102].

Currently, there are many small humanoid robots for research and hobby use. For example, we can choose NAO by Aldebaran Robotics [1], DARwIn-OP by ROBOTIS [103], PALRO by FujitSoft [33], or KHR series by Kondo Kagaku Co. Ltd. [4]

DARPA Robotics Challenge

On April 10, 2012, the Defense Advanced Research Projects Agency (DARPA) of the United States announced a program , namely, the DARPA Robotics Challenge (DRC) [3]. It's primary goal is to develop robotics technologies which can manage complex tasks in dangerous, degraded, and human engineered environment by utilizing available human hand tools, devices, and vehicles [5]. DRC is a competition style project where many teams (robots) compete their performances on the same task. In the trial of December 2013, the following tasks were specified.

1. Drive a utility vehicle
2. Travel dismounted

3. Remove debris blocking entry
4. Open a door, enter building
5. Climb an industrial ladder
6. Break through a wall
7. Locate and close valves
8. Carry, unspool, and connect a fire hose

Note that DRC is not limited to the robot configuration being humanoid, but they are expecting human-like competence for the given tasks. Indeed some teams have designed non-humanoid robots like CHIMP by Carnegie Mellon University (CMU) – National robotics Engineering Center (NREC) [2] and ROBOSIMIAN by NASA – Jet Propulsion Laboratory [45]. Yet however, dominant participant teams have chosen humanoid robot designs for this challenge. Moreover, DRC offers an official humanoid robot Atlas developed by Boston Dynamics. Its copies will be used by seven teams. The DRC final will be held in December 2014, and with no doubt, the DARPA Robotics Challenge will have an enormous impact to humanoid robotics research in the world.

Chapter 2
Kinematics

The theory to analyze the relationship between the position and attitude of a link and the joint angles of a mechanism is called **Kinematics**. It is the basis on which robotics is formed, but the exact same theory is used for computer graphics as well. Both require mathematics and algorithms which can clearly represent a moving object in 3D space.

2.1 Coordinate Transformations

Figure 2.1 illustrates the Humanoid Robot HRP-2, which was developed during the Humanoid Robotics Project [65]. It has a height of 154 cm, and its weight including the batteries is 58 kg. It can walk for about one hour using the batteries alone. The HRP-2 has a total of 30 joints which can be controlled independently. Fig 2.1(b) shows the names of the Links and the position of their Local Coordinates.

2.1.1 World Coordinates

Our first task is to clearly define the position of each part of the robot. To do this we define a special point directly below the robot in its initial position[1] as shown in Fig 2.2. Taking this point as the origin we define a fixed coordinate system whose x axis faces forward, y axis to the left and z axis faces up.

We will name these coordinates Σ_W and hereafter will use this to describe our robots and their motion. Coordinates such as this one are called **World Coordinates**. By using a common coordinate system to define the robot itself and surrounding objects we can check a safe landing of a foot on the ground, successful hand grasping of an object, undesired collisions with the environment, and so on.

[1] Actually it is the intersection between a perpendicular line which goes through the origin of the the Local Coordinates of the Waist Link and the floor.

S. Kajita et al., *Introduction to Humanoid Robotics*,
Springer Tracts in Advanced Robotics 101,
DOI: 10.1007/978-3-642-54536-8_2, © Springer-Verlag Berlin Heidelberg 2014

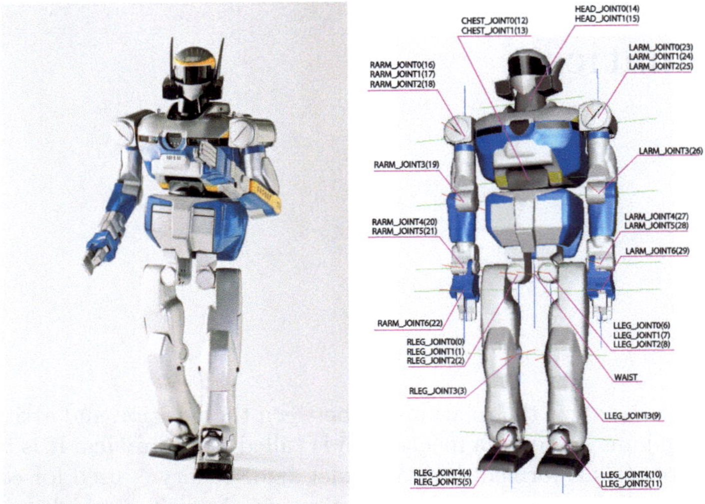

(a) (b)

Fig. 2.1 (a) The Humanoid Robot HRP-2. (b) All Links have Names and Local Coordinates.

Positions defined using World Coordinates are called **Absolute Positions**. For example, we describe the absolute position of the left hand tip in Fig. 2.2 by the following 3D vectors

$$\boldsymbol{p}_h = \begin{bmatrix} p_{hx} \\ p_{hy} \\ p_{hz} \end{bmatrix}.$$

We will also use the terms, **Absolute Attitude** and **Absolute Velocity** to show they are defined in the World Coordinates.

2.1.2 Local Coordinates and Homogeneous Transformations

Lets take a look at how the hand tip position $\boldsymbol{p}_h$ changes by the rotation of the robot's shoulder. From Fig. 2.3(a) we can see that the absolute position of the left shoulder is defined by vector $\boldsymbol{p}_a$, and vector $\boldsymbol{r}$ shows the position of the hand end relative to the shoulder. As we can see from this figure, we have

$$\boldsymbol{p}_h = \boldsymbol{p}_a + \boldsymbol{r}.$$

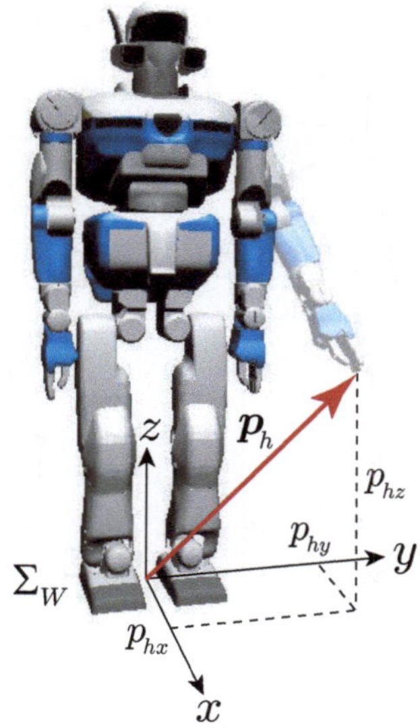

Fig. 2.2 Origin of World Coordinates defined directly below robot in initial position. $\boldsymbol{p}_h$ is the location of the end of the hands in World Coordinates.

At the open arm pose of Fig. 2.3(b), the absolute hand end position can be expressed by introducing another vector $\boldsymbol{r}'$ which points the lifted hand from the shoulder.

$$\boldsymbol{p}_h = \boldsymbol{p}_a + \boldsymbol{r}'. \tag{2.1}$$

The hand lifts by the vector's rotation from $\boldsymbol{r}$ to $\boldsymbol{r}'$ while the shoulder keeps the same position $\boldsymbol{p}_a$.

Let us think of a **local coordinate system** Σ_a which is attached to the left shoulder as shown in Fig. 2.3. Unlike the world coordinate system which is fixed to the ground, the local coordinates "move" together with the attached link.

The local coordinate system Σ_a is defined by the origin at the shoulder and the unit vectors that describe the x, y and z axes. The three unit vectors $\boldsymbol{e}_{ax}$, $\boldsymbol{e}_{ay}$ and $\boldsymbol{e}_{az}$ are defined to be parallel to Σ_W at the initial arm down pose as shown in Fig. 2.3(a). When the robot lifts its arm, Σ_a rotates about axis $\boldsymbol{e}_{ax}$ with the amount of angle ϕ as illustrated in Fig. 2.3(b).

The relationship between ϕ, the shoulder rotation, and Σ_a can be described by the following equations

$$\boldsymbol{e}_{ax} = \begin{bmatrix} 1 \\ 0 \\ 0 \end{bmatrix}, \quad \boldsymbol{e}_{ay} = \begin{bmatrix} 0 \\ \cos\phi \\ \sin\phi \end{bmatrix}, \quad \boldsymbol{e}_{az} = \begin{bmatrix} 0 \\ -\sin\phi \\ \cos\phi \end{bmatrix}. \tag{2.2}$$

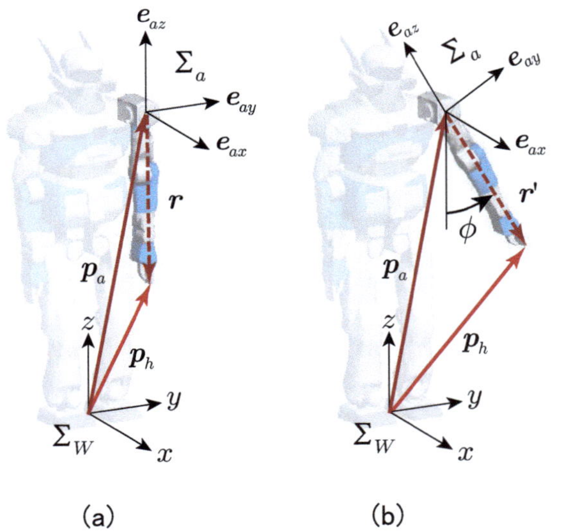

Fig. 2.3 World Coordinates Σ_W and Local Coordinates of Arm Σ_a. (a) Initial Position: Σ_W and Σ_a are parallel. (b) Σ_a rotates together with Arm.

Since it is a rotation around the x axis, only $\boldsymbol{e}_{ay}$ and $\boldsymbol{e}_{az}$ change by ϕ. Let us define a 3×3 matrix, $\boldsymbol{R}_a$ as follows:

$$\boldsymbol{R}_a \equiv [\boldsymbol{e}_{ax}\, \boldsymbol{e}_{ay}\, \boldsymbol{e}_{az}]. \tag{2.3}$$

Using the matrix $\boldsymbol{R}_a$ we can describe the relationship between $\boldsymbol{r}$ and $\boldsymbol{r}'$ in Fig. 2.3 as follows:

$$\boldsymbol{r}' = \boldsymbol{R}_a \boldsymbol{r}. \tag{2.4}$$

In other words, a vector is rotated by multiplying it by the matrix $\boldsymbol{R}_a$. We will go over this in more detail in Section 2.2.

Now, let us define the position of the end of the hand in the local coordinate system Σ_a as ${}^a\boldsymbol{p}_h$. The a at the top left means that it is in the local coordinate system defined by Σ_a. From Fig. 2.3(a) we get,

$${}^a\boldsymbol{p}_h = \boldsymbol{r}. \tag{2.5}$$

In the example shown in Fig. 2.3, Σ_a and the left arm move as one block, so vector ${}^a\boldsymbol{p}_h$ remains constant.

To describe the position of the end of the left arm, we have used two different notations.

- Position of end of hand $\boldsymbol{p}_h$ viewed from from the world coordinate Σ_W.
- Position of end of hand ${}^a\boldsymbol{p}_h$ viewed from the local coordinate Σ_a.

The relationship between the two are as follows. From (2.1), (2.4) and (2.5), we get,

$$\boldsymbol{p}_h = \boldsymbol{p}_a + \boldsymbol{R}_a\, {}^a\boldsymbol{p}_h. \tag{2.6}$$

Equation (2.6) can also be rewritten as,

$$\begin{bmatrix} \boldsymbol{p}_h \\ 1 \end{bmatrix} = \begin{bmatrix} \boldsymbol{R}_a & \boldsymbol{p}_a \\ 0\,0\,0 & 1 \end{bmatrix} \begin{bmatrix} {}^a\boldsymbol{p}_h \\ 1 \end{bmatrix}. \tag{2.7}$$

Here we added 0 and 1 to the matrices to match the dimensions and make it match (2.6). The 4×4 matrix on the right hand side of the equation comes from combining the position vector $\boldsymbol{p}_a$ and the rotation matrix $\boldsymbol{R}_a$. This can be written as,

$$\boldsymbol{T}_a \equiv \begin{bmatrix} \boldsymbol{R}_a & \boldsymbol{p}_a \\ 0\,0\,0 & 1 \end{bmatrix}.$$

Matrices like this are called **Homogeneous transformation matrices**[2]. The Homogeneous Transformation $\boldsymbol{T}_a$ converts points described in arm local coordinates to world coordinates

$$\begin{bmatrix} \boldsymbol{p} \\ 1 \end{bmatrix} = \boldsymbol{T}_a \begin{bmatrix} {}^a\boldsymbol{p} \\ 1 \end{bmatrix}.$$

Note that the point ${}^a\boldsymbol{p}$ can be any point in the left arm, thus the arm shape can be represented by its aggregation. On the other hand, the arm configuration (position and orientation) is separately represented by the matrix $\boldsymbol{T}_a$. In general, we can say that a **Homogeneous Transformation in itself describes the position and attitude of an object.**

2.1.3 Local Coordinate Systems Local to Local Coordinate Systems

It is possible to define a Local Coordinate System which has another Local Coordinate System as it's parent. In Fig. 2.4(a) we show a Local Coordinate System Σ_b which has Σ_a as it's parent. Σ_b moves together with the lower arm and is set so that the axes are in-line with the axes of Σ_a when the elbow is straight. Let us define the rotation angle of the elbow joint as θ. Let us also define the unit vectors of which are the axes x, y, z of Σ_b as ${}^a\boldsymbol{e}_{bx}, {}^a\boldsymbol{e}_{by}, {}^a\boldsymbol{e}_{bz}$ respectively. They are defined as follows,

[2] The same matrix is referred to in computer graphics as an **Affine transformation matrix**. There is a difference as in an Affine transformation, $\boldsymbol{R}_a$ is not necessarily a simple rotation matrix (→ Section 2.2) and may include scaling and shearing as well.

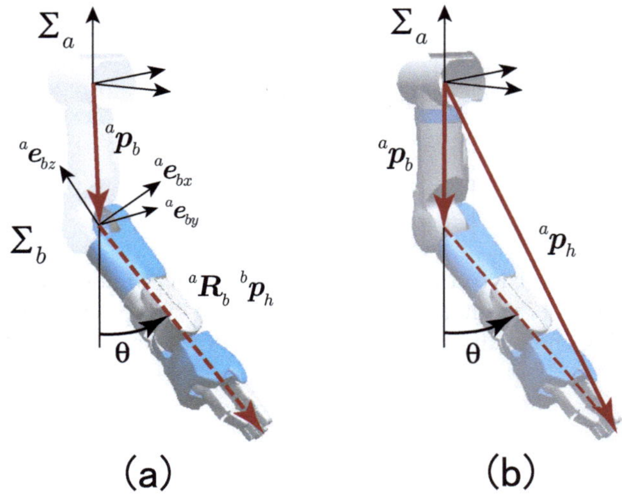

Fig. 2.4 Arm Local Coordinates Σ_a and the Lower Arm Local Coordinates Σ_b

$$ {}^a\boldsymbol{e}_{bx} = \begin{bmatrix} \cos\theta \\ 0 \\ \sin\theta \end{bmatrix}, \quad {}^a\boldsymbol{e}_{by} = \begin{bmatrix} 0 \\ 1 \\ 0 \end{bmatrix}, \quad {}^a\boldsymbol{e}_{bz} = \begin{bmatrix} -\sin\theta \\ 0 \\ \cos\theta \end{bmatrix}. \tag{2.8} $$

Since the elbow rotates around the y axis, only the vectors ${}^a\boldsymbol{e}_{bx}$ and ${}^a\boldsymbol{e}_{bz}$ change by θ. The vectors are defined in Σ_a space and thus they have the indicator a on the top left. Let us define a matrix ${}^a\boldsymbol{R}_b$ as a combination of the three unit vectors.

$$ {}^a\boldsymbol{R}_b \equiv [{}^a\boldsymbol{e}_{bx}\ {}^a\boldsymbol{e}_{by}\ {}^a\boldsymbol{e}_{bz}] \tag{2.9} $$

The conversion of ${}^b\boldsymbol{p}_h$ (in Fig. 2.4(a)) which is in Σ_b space, to ${}^a\boldsymbol{p}_h$ (in Fig. 2.4 (b)) in Σ_a space becomes as follows

$$ \begin{bmatrix} {}^a\boldsymbol{p}_h \\ 1 \end{bmatrix} = {}^a\boldsymbol{T}_b \begin{bmatrix} {}^b\boldsymbol{p}_h \\ 1 \end{bmatrix}. \tag{2.10} $$

Here we defined a homogeneous transformation ${}^a\boldsymbol{T}_b$ as follows

$$ {}^a\boldsymbol{T}_b \equiv \begin{bmatrix} {}^a\boldsymbol{R}_b & {}^a\boldsymbol{p}_b \\ 0\,0\,0 & 1 \end{bmatrix}. $$

Note that ${}^a\boldsymbol{p}_b$ is the origin of Σ_b viewed from Σ_a.

By combining the results of (2.10) with (2.7) we get the equation which converts a point defined in Σ_b space to world coordinates. The example below converts the manipulator end point in Σ_b

$$ \begin{bmatrix} \boldsymbol{p}_h \\ 1 \end{bmatrix} = \boldsymbol{T}_a\, {}^a\boldsymbol{T}_b \begin{bmatrix} {}^b\boldsymbol{p}_h \\ 1 \end{bmatrix}. \tag{2.11} $$

Let us define the product of homogeneous transformation on the right hand side of the equation as one matrix and call it $\boldsymbol{T}_b$

$$\boldsymbol{T}_b \equiv \boldsymbol{T}_a {}^a\boldsymbol{T}_b.$$

The matrix $\boldsymbol{T}_b$ is a Homogeneous Transformation which is in fact Σ_b described in World Coordinates, and it describes the position and attitude of the forearm. $\boldsymbol{T}_a$ describes the amount the shoulder has been turned. ${}^a\boldsymbol{T}_b$ changes together with the amount the elbow is rotated. Therefore we can see that the description of $\boldsymbol{T}_b$ shows the effect of the shoulder and the elbow rotations.

2.1.4 Homogeneous Transformations and Chain Rules

What we described above can be generalised. Let us assume we have a mechanism which connects the local coordinates Σ_1 through Σ_N. If we have a Homogeneous Transformation which describes neighbouring local coordinates Σ_i and Σ_{i+1} such as,

$${}^i\boldsymbol{T}_{i+1},$$

by reiterating through the above process we get the following equation

$$\boldsymbol{T}_N = \boldsymbol{T}_1 {}^1\boldsymbol{T}_2 {}^2\boldsymbol{T}_3 \ldots {}^{N-1}\boldsymbol{T}_N. \tag{2.12}$$

Here $\boldsymbol{T}_N$ is a homogeneous transformation that describes the position and attitude of the Nth joint. To add another link to the end of this joint we need to multiply it with the homogeneous transformation matrix **from the right**.

This method of multiplying the homogeneous transformation in order to calculate the coordinate transform matrix is called the **chain rule** [87]. The chain rule enables us to calculate the kinematics of an arm with multiple joints without too much complication.

2.2 Characteristics of Rotational Motion

In the previous section we described the rotation of the robot's joint with a 3×3 matrix with no explanation. This matrix is called a **Rotation Matrix**. It describes the attitude of the link and also notates the rotational motion. In this section we will go over the characteristics of rotation in three dimensions using the Rotation Matrix. To simplify we will limit the scope to rotation around the origin.

2.2.1 Roll, Pitch and Yaw Notation

The basics of rotation are the rotations around the x, y and z axes, which we will call Roll, Pitch and Yaw respectively. In Fig. 2.5 we show a triangle that resides on the xy plane going through Roll, Pitch and Yaw.

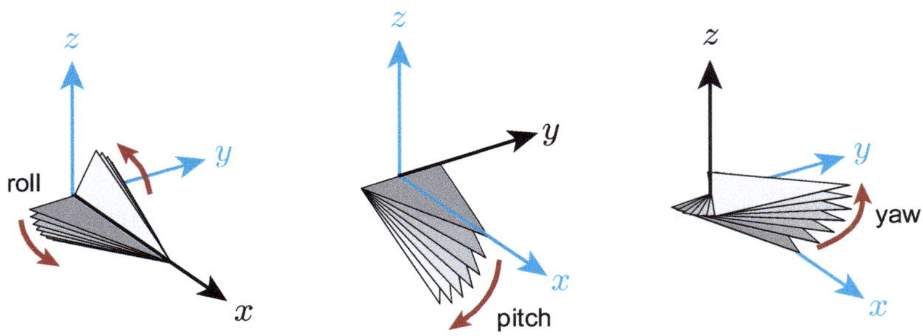

Fig. 2.5 Roll, Pitch and Yaw using the x, y and z axes. Figure shows the triangle rotation at every $\pi/18$ up to $+\pi/3$ [rad]. If we think the triangle is a plane flying along the x axis, they correspond to a right bank, nose down and left turn.

Below is a list of rotation axes, their names and often used notations.

Rotation Axis	Name	Notation
x axis	Roll	ϕ
y axis	Pitch	θ
z axis	Yaw	ψ

To Roll, Pitch and Yaw an object a given angle we need to use the following matrices

$$\boldsymbol{R}_x(\phi) = \begin{bmatrix} 1 & 0 & 0 \\ 0 & \cos\phi & -\sin\phi \\ 0 & \sin\phi & \cos\phi \end{bmatrix}$$

$$\boldsymbol{R}_y(\theta) = \begin{bmatrix} \cos\theta & 0 & \sin\theta \\ 0 & 1 & 0 \\ -\sin\theta & 0 & \cos\theta \end{bmatrix}$$

$$\boldsymbol{R}_z(\psi) = \begin{bmatrix} \cos\psi & -\sin\psi & 0 \\ \sin\psi & \cos\psi & 0 \\ 0 & 0 & 1 \end{bmatrix}.$$

If we Roll, Pitch and then Yaw a point $\boldsymbol{p}$ around the origin, it will move to the point,

$$\boldsymbol{p}' = \boldsymbol{R}_z(\psi)\boldsymbol{R}_y(\theta)\boldsymbol{R}_x(\phi)\,\boldsymbol{p}.$$

We can rewrite this as

$$\boldsymbol{p}' = \boldsymbol{R}_{rpy}(\phi, \theta, \psi)\, \boldsymbol{p}.$$

Here we have introduced the following notation

$$\begin{aligned}\boldsymbol{R}_{rpy}(\phi, \theta, \psi) &\equiv \boldsymbol{R}_z(\psi)\boldsymbol{R}_y(\theta)\boldsymbol{R}_x(\phi) \\ &= \begin{bmatrix} c_\psi c_\theta & -s_\psi c_\phi + c_\psi s_\theta s_\phi & s_\psi s_\phi + c_\psi s_\theta c_\phi \\ s_\psi c_\theta & c_\psi c_\phi + s_\psi s_\theta s_\phi & -c_\psi s_\phi + s_\psi s_\theta c_\phi \\ -s_\theta & c_\theta s_\phi & c_\theta c_\phi \end{bmatrix}, \end{aligned} \tag{2.13}$$

where $c_\psi \equiv \cos\psi$, $s_\psi \equiv \sin\psi$, $c_\theta \equiv \cos\theta$, and so on.

The matrix $\boldsymbol{R}_{rpy}(\phi, \theta, \psi)$ is also a rotation matrix and it is known that any given attitude in 3D space can be realized by using this. Therefore, arbitrary attitude can be represented just by a set of three numbers (ϕ, θ, ψ), which are called the Roll-Pitch-Yaw notation or the Z-Y-X Euler angles.

Since the Roll-Pitch-Yaw notation can be easily understood, it is frequently used to notate attitudes of ships, airplanes, and robots.

2.2.2 The Meaning of Rotation Matrices

There are two different ways to interpret rotation matrices. One is as **an operator that rotates vectors**. Figure 2.6(a) shows a vector $\boldsymbol{p}$ rotated by the rotation matrix $\boldsymbol{R}$ to a new vector $\boldsymbol{p}'$

$$\boldsymbol{p}' = \boldsymbol{R}\boldsymbol{p}$$

where $\boldsymbol{p}$ and $\boldsymbol{p}'$ reside in the same coordinate space.

The second way to interpret $\boldsymbol{R}$ is as **the attitude of a local coordinate space**. Figure 2.6(b) shows local coordinate axes $(\boldsymbol{e}_x, \boldsymbol{e}_y, \boldsymbol{e}_z)$ and a point $\bar{\boldsymbol{p}}$

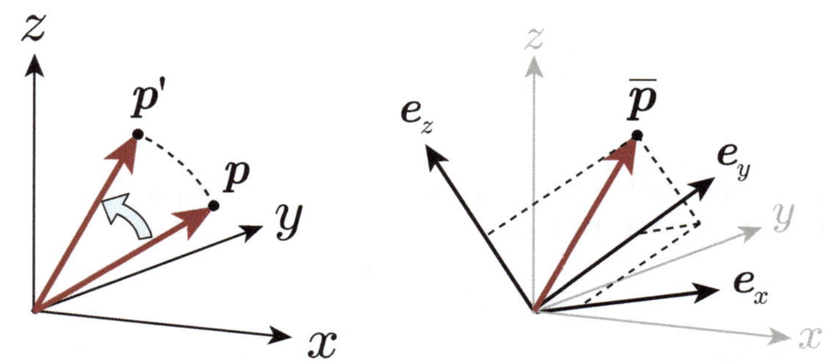

(a) Operator to rotate a vector (b) Attitude of a local frame

Fig. 2.6 Two ways to interpret rotation matrices

defined in this local coordinates. The rotation matrix is defined as the set of the coordinate axes, such that

$$\boldsymbol{R} \equiv [\boldsymbol{e}_x \ \boldsymbol{e}_y \ \boldsymbol{e}_z]. \tag{2.14}$$

The point is represented in world coordinates as

$$\boldsymbol{p} = \boldsymbol{R}\bar{\boldsymbol{p}}.$$

In this case, we merely changed the point of view on a stationary point without any actual motion.

How can we figure out a rotation matrix used in which meaning, (1) operator to rotate a vector, or (2) attitude of a local frame? We can see it by checking whether the point exists in the same coordinate space before and after the calculation. Most of the time, however, it is clear from the context.

2.2.3 Calculating the Inverse of a Rotation Matrix

Let as assume that we have a rotation matrix showing attitude of a local frame. We will define the unit vectors of this coordinate system as $\boldsymbol{e}_x$ $\boldsymbol{e}_y$ and $\boldsymbol{e}_z$. As these unit vectors are at right angles to each other we can say,

$$\boldsymbol{e}_i^T \boldsymbol{e}_j = \begin{cases} 1 \ (i = j) \\ 0 \ (i \neq j) \end{cases}.$$

Here $\boldsymbol{e}_i^T \boldsymbol{e}_j$ is the inner product of two vectors. If we calculate product of the transpose $\boldsymbol{R}^T$ and $\boldsymbol{R}$ we get,

$$\begin{aligned} \boldsymbol{R}^T \boldsymbol{R} &= \begin{bmatrix} \boldsymbol{e}_x^T \\ \boldsymbol{e}_y^T \\ \boldsymbol{e}_z^T \end{bmatrix} [\boldsymbol{e}_x \ \boldsymbol{e}_y \ \boldsymbol{e}_z] = \begin{bmatrix} \boldsymbol{e}_x^T \boldsymbol{e}_x & \boldsymbol{e}_x^T \boldsymbol{e}_y & \boldsymbol{e}_x^T \boldsymbol{e}_z \\ \boldsymbol{e}_y^T \boldsymbol{e}_x & \boldsymbol{e}_y^T \boldsymbol{e}_y & \boldsymbol{e}_y^T \boldsymbol{e}_z \\ \boldsymbol{e}_z^T \boldsymbol{e}_x & \boldsymbol{e}_z^T \boldsymbol{e}_y & \boldsymbol{e}_z^T \boldsymbol{e}_z \end{bmatrix} \\ &= \begin{bmatrix} 1 & 0 & 0 \\ 0 & 1 & 0 \\ 0 & 0 & 1 \end{bmatrix} = \boldsymbol{E} \ (3 \times 3 \text{ identity matrix}). \end{aligned}$$

From this we can safely say that $\boldsymbol{R}^T \boldsymbol{R} = \boldsymbol{E}$. If we multiply both sides of this equation by $\boldsymbol{R}^{-1}$ from the right,

$$\boldsymbol{R}^T = \boldsymbol{R}^{-1}. \tag{2.15}$$

Therefore, when we transpose a rotation matrix we get its inverse. Matrices which have this characteristic are called **orthogonal matrices**.

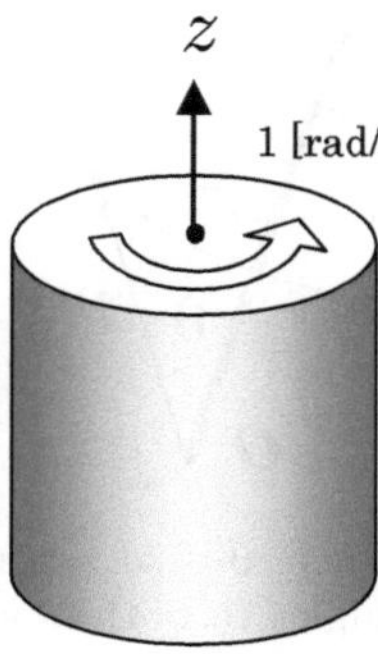

Fig. 2.7 A Rotating Cylinder. We define its angular velocity vector to be $[0\ 0\ 1]^T$ [rad/s].

2.2.4 Angular Velocity Vector

Next we will define a way to notate rotational speed in three dimensional space. The simplest example is shown in Fig. 2.7. This figure shows a cylinder rotating around the z axis at 1 [rad/s]. In this case the rotational velocity of this object, $\boldsymbol{\omega}$, can be described by

$$\boldsymbol{\omega} = \begin{bmatrix} 0 \\ 0 \\ 1 \end{bmatrix}. \tag{2.16}$$

This is called an **Angular Velocity Vector**. Each element in this vector has the unit of [rad/s].

An Angular Velocity Vector($\boldsymbol{\omega}$) has the following characteristics.

1 **$\boldsymbol{\omega}$ is defined as a unit vector × scalar value**

Let $\boldsymbol{a}$ be a unit vector on the rotating axis and $\dot{q}$ be the rotational speed (a scalar value). Then the angular velocity vector of the object is their product

$$\boldsymbol{\omega} = \boldsymbol{a}\dot{q}. \tag{2.17}$$

2 **$\boldsymbol{\omega}$ describes the velocity of all the points on the rotating object**

Let $\boldsymbol{p}$ be a vector which represents a given point p on the surface of the object. Let us also define the base of this vector to be somewhere on the rotation axis. In this case the velocity of point p will be described by $\boldsymbol{\omega} \times \boldsymbol{p}$. The × operator is the **cross product** or **outer product** of two vectors and is defined as follows.

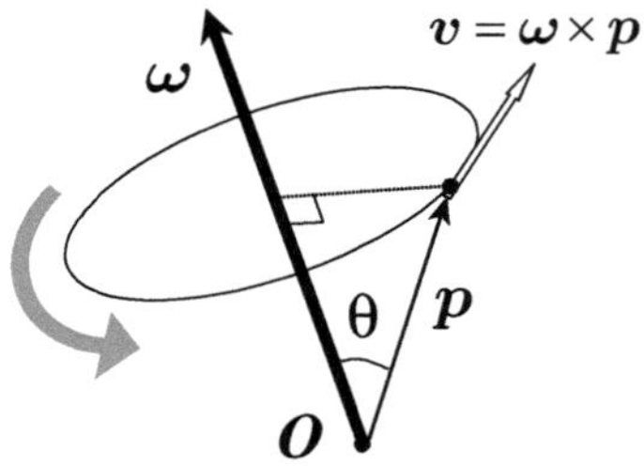

Fig. 2.8 Definition of Cross Product. $\boldsymbol{v} = \boldsymbol{\omega} \times \boldsymbol{p}$ represents the velocity of the a point on the circle. It is at right angles to both $\boldsymbol{\omega}$ and $\boldsymbol{p}$.

By using the two vectors, $\boldsymbol{\omega}$ and $\boldsymbol{p}$ we obtain a new vector $\boldsymbol{v}$ which has the following characteristics (Fig. 2.8)

$$|\boldsymbol{v}| = |\boldsymbol{\omega}||\boldsymbol{p}| \sin\theta$$
$$(\boldsymbol{v} \perp \boldsymbol{\omega}) \cap (\boldsymbol{v} \perp \boldsymbol{p}).$$

However, there are two vectors that satisfy this condition so we use the right screw rule and choose the one which matches the direction a right handed screw will move when turned. This can be represented as

$$\boldsymbol{v} = \boldsymbol{\omega} \times \boldsymbol{p} \tag{2.18}$$

and this is called the "cross product of $\boldsymbol{\omega}$ and $\boldsymbol{p}$".

When given the elements of $\boldsymbol{\omega}$ and $\boldsymbol{p}$, the cross product can be calculated as follows. As an exercise we can also check to see that the equations below can be derived from the above.

$$\boldsymbol{\omega} \times \boldsymbol{p} = \begin{bmatrix} \omega_x \\ \omega_y \\ \omega_z \end{bmatrix} \times \begin{bmatrix} p_x \\ p_y \\ p_z \end{bmatrix} \equiv \begin{bmatrix} \omega_y p_z - \omega_z p_y \\ \omega_z p_x - \omega_x p_z \\ \omega_x p_y - \omega_y p_x \end{bmatrix} \tag{2.19}$$

We can say that the cross product has been defined so that it matches the nature of angular velocity[3]. The equation below adds "Physical Meaning" to the angular velocity vector.

$$\text{(Velocity of Vertex)} = \boldsymbol{\omega} \times \text{(Position of Vertex)}$$

We have used the equations above to visualise the surface velocity of a rugby ball shaped object (an ellipsoid) rotating at $\boldsymbol{\omega}$ as shown in Fig. 2.9. On

[3] From this point on we will use this to calculate moment and angular momentum. The cross product itself is important in electromagnetism. Flemming's Right Hand Rule and Left Hand Rule are themselves the result of cross products.

a rotating object, the velocity can have different magnitudes and directions depending the location of the point on the surface. The angular velocity vector represents this fact by using just three elements[4].

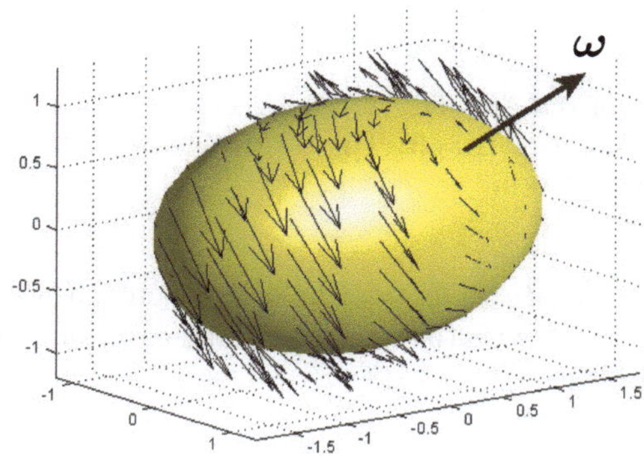

Fig. 2.9 Surface Velocity on an Ellipsoid. The object rotates according to angular velocity vector $\boldsymbol{\omega}$ (Thick Arrow). The resulting surface velocity is represented in small arrows.

3 ω can be rotated too

Let us multiply both sides of (2.17) using some rotation matrix $\boldsymbol{R}$

$$\boldsymbol{R}\boldsymbol{\omega} = \boldsymbol{R}\boldsymbol{a}\dot{q}. \tag{2.20}$$

If we introduce following new vectors,

$$\boldsymbol{\omega}' = \boldsymbol{R}\boldsymbol{\omega}, \quad \boldsymbol{a}' = \boldsymbol{R}\boldsymbol{a}$$

then we can rewrite (2.20) as

$$\boldsymbol{\omega}' = \boldsymbol{a}'\dot{q}.$$

This matches the definition in (2.17), so we can see that $\boldsymbol{\omega}'$ is a new angular velocity vector around the axis defined by $\boldsymbol{a}'$. Therefore we can say that an angular velocity vector can be transformed directly by using a rotation matrix $\boldsymbol{R}$.

[4] If you want strict mathematical facts, the angular velocity vector is called a pseudovector (or axial vector), and it is treated a little differently from normal vectors [105]. However within the scope of this book there is no problem in treating it as a normal vector. In this sense, the torque and angular momentum are also represented using pseudo-vectors as well.

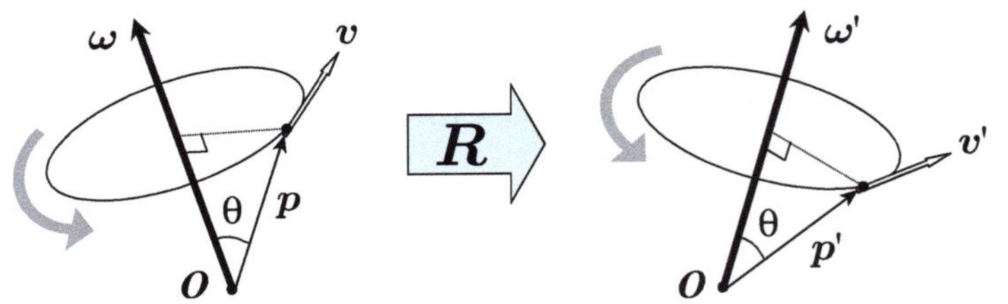

Fig. 2.10 Rotation of the Angular Velocity Vector, Position Vector and Velocity Vector using the rotation matrix $\boldsymbol{R}$. The relation between the three vectors does not change after the tranform.

Next let us try to rotate an angular velocity vector, a position vector, and the corresponding velocity vector using the same rotation matrix $\boldsymbol{R}$ as shown in Fig. 2.10 which gives

$$\boldsymbol{\omega}' = \boldsymbol{R}\boldsymbol{\omega}, \quad \boldsymbol{p}' = \boldsymbol{R}\boldsymbol{p}, \quad \boldsymbol{v}' = \boldsymbol{R}\boldsymbol{v}.$$

Since the spatial relation between these three vectors does not change by this rotation, the definition of the cross product must be preserved, so

$$\boldsymbol{v}' = \boldsymbol{\omega}' \times \boldsymbol{p}'. \tag{2.21}$$

Substituting the rotated vectors $\boldsymbol{v}', \boldsymbol{\omega}'$ and $\boldsymbol{p}'$ by their originals $\boldsymbol{R}\boldsymbol{v}, \boldsymbol{R}\boldsymbol{\omega}$ and $\boldsymbol{R}\boldsymbol{p}$, we have

$$\boldsymbol{R}(\boldsymbol{\omega} \times \boldsymbol{p}) = (\boldsymbol{R}\boldsymbol{\omega}) \times (\boldsymbol{R}\boldsymbol{p}). \tag{2.22}$$

2.2.5 Differentiation of the Rotation Matrix and Angular Velocity Vectors

Now let us look into the relationship between the angular velocity vector, $\boldsymbol{\omega}$ and the rotation matrix, $\boldsymbol{R}$. In Section 2.2.2 we saw that the rotation matrix $\boldsymbol{R}$ gives the relationship between the local coordinates and world coordinates of a vertex

$$\boldsymbol{p} = \boldsymbol{R}\bar{\boldsymbol{p}}. \tag{2.23}$$

If we differentiate this equation with respect to time we get the velocity of the point in the world coordinate space. The position of the point, $\bar{\boldsymbol{p}}$ does not move in the local coordinate space so,

$$\dot{\boldsymbol{p}} = \dot{\boldsymbol{R}}\bar{\boldsymbol{p}}. \tag{2.24}$$

Here we substitute $\bar{\boldsymbol{p}}$ using the relationship $\bar{\boldsymbol{p}} = \boldsymbol{R}^T\boldsymbol{p}$ to get

$$\dot{\boldsymbol{p}} = \dot{\boldsymbol{R}}\boldsymbol{R}^T\boldsymbol{p}. \tag{2.25}$$

We can use this equation to calculate the velocity of a vertex on the object.

From the previous section we know that $\dot{\boldsymbol{p}} = \boldsymbol{\omega} \times \boldsymbol{p}$ so we can safely state the following relationship

$$\boldsymbol{\omega} \times \boldsymbol{p} = \dot{\boldsymbol{R}}\boldsymbol{R}^T\boldsymbol{p}. \tag{2.26}$$

If we remember that the cross product is calculated as follows (2.19),

$$\boldsymbol{\omega} \times \boldsymbol{p} = \begin{bmatrix} \omega_x \\ \omega_y \\ \omega_z \end{bmatrix} \times \begin{bmatrix} p_x \\ p_y \\ p_z \end{bmatrix} = \begin{bmatrix} \omega_y p_z - \omega_z p_y \\ \omega_z p_x - \omega_x p_z \\ \omega_x p_y - \omega_y p_x \end{bmatrix}. \tag{2.27}$$

We can rewrite this as a product of a 3×3 matrix to obtain.

$$\boldsymbol{\omega} \times \boldsymbol{p} = \begin{bmatrix} 0 & -\omega_z & \omega_y \\ \omega_z & 0 & -\omega_x \\ -\omega_y & \omega_x & 0 \end{bmatrix} \begin{bmatrix} p_x \\ p_y \\ p_z \end{bmatrix} \equiv \boldsymbol{S}\boldsymbol{p}. \tag{2.28}$$

The matrix we just defined has an interesting characteristic. If we take a transpose of this matrix we get the same matrix with only the sign inverted. Matrices such as this are called **skew symmetric matrices** and satisfy

$$\boldsymbol{S}^T = -\boldsymbol{S}. \tag{2.29}$$

By comparing (2.26) and (2.28) we can see that $\dot{\boldsymbol{R}}\boldsymbol{R}^T$ is in fact a Skew Symmetric Matrix. This can be proved as follows. First the transpose of a rotation matrix also happens to be it's inverse matrix. So,

$$\boldsymbol{R}\boldsymbol{R}^T = \boldsymbol{E} \tag{2.30}$$

Furthermore, if we take the time derivative of this equation[5] we get the following.

$$\begin{aligned} \dot{\boldsymbol{R}}\boldsymbol{R}^T + \boldsymbol{R}\dot{\boldsymbol{R}}^T &= \boldsymbol{0} \\ (\dot{\boldsymbol{R}}\boldsymbol{R}^T)^T &= -\dot{\boldsymbol{R}}\boldsymbol{R}^T \end{aligned}$$

Which proves that $\dot{\boldsymbol{R}}\boldsymbol{R}^T$ is in fact a Skew Symmetric Matrix.

In this book we will refer to 3 dimensional vectors taken from Skew Symmetric Matrices as the $\vee$ "wedge" operation. Making a Skew Symmetric Matrix from a 3 dimensional vector will be referred to as taking the $\wedge$ "hat" operation as follows

[5] The derivative of a matrix is to simply take the derivative of all it's components. We must take care not to change the order of the multiplication.

$$\begin{bmatrix} 0 & -\omega_z & \omega_y \\ \omega_z & 0 & -\omega_x \\ -\omega_y & \omega_x & 0 \end{bmatrix}^{\vee} = \begin{bmatrix} \omega_x \\ \omega_y \\ \omega_z \end{bmatrix}$$

$$\begin{bmatrix} \omega_x \\ \omega_y \\ \omega_z \end{bmatrix}^{\wedge} = \begin{bmatrix} 0 & -\omega_z & \omega_y \\ \omega_z & 0 & -\omega_x \\ -\omega_y & \omega_x & 0 \end{bmatrix}.$$

Therefore a cross product can be written as follows.

$$\boldsymbol{\omega} \times \boldsymbol{p} = (\boldsymbol{\omega}^{\wedge})\boldsymbol{p}$$

To make this equation easier on the eyes we can put the $\wedge$ at the top.[6]

$$\boldsymbol{\omega} \times \boldsymbol{p} = \widehat{\boldsymbol{\omega}}\boldsymbol{p}$$

Using these methods of description we can describe the relationship between a rotation matrix and its angular velocity in (2.26) by

$$\widehat{\boldsymbol{\omega}} = \dot{\boldsymbol{R}}\boldsymbol{R}^T \tag{2.31}$$

or

$$\boldsymbol{\omega} = (\dot{\boldsymbol{R}}\boldsymbol{R}^T)^{\vee}. \tag{2.32}$$

2.2.6 Integration of the Angular Velocity Vector and Matrix Exponential

Lets take a look at how we can integrate an angular velocity vector and get the rotation matrix. By multiplying both sides of (2.31) by $\boldsymbol{R}$ from the right we obtain

$$\dot{\boldsymbol{R}} = \widehat{\boldsymbol{\omega}}\boldsymbol{R}. \tag{2.33}$$

This equation is important because this gives the relationship between angular velocity vectors and rotation matrices. We will call this equation the **Basic Equation of Rotation**. This equation is a differential equation for the matrix $\boldsymbol{R}$ so if you integrate this you get a rotation matrix. If the initial condition is $\boldsymbol{R}(0) = \boldsymbol{E}$ and the angular velocity $\boldsymbol{\omega}$ is constant, the solution is:

$$\boldsymbol{R}(t) = \boldsymbol{E} + \widehat{\boldsymbol{\omega}}t + \frac{(\widehat{\boldsymbol{\omega}}t)^2}{2!} + \frac{(\widehat{\boldsymbol{\omega}}t)^3}{3!} + \ldots \quad . \tag{2.34}$$

By substituting the above into (2.33) we can confirm that this is correct. Using the analogy with exponential function we will describe the solution using $e^{\widehat{\omega}t}$ and will call them **matrix exponential**. Therefore we define,

[6] A lot of text books and papers use $[\boldsymbol{\omega}\times]$ instead of $\widehat{\boldsymbol{\omega}}$ to put an emphasis on the fact that this is the cross product.

$$e^{\widehat{\omega}t} \equiv \boldsymbol{E} + \widehat{\omega}t + \frac{(\widehat{\omega}t)^2}{2!} + \frac{(\widehat{\omega}t)^3}{3!} + \ldots. \tag{2.35}$$

This infinite series can be simplified [101]. First we decompose $\boldsymbol{\omega}$ into a unit vector $\boldsymbol{a}$ and a scalar ω.

$$\boldsymbol{\omega} = \boldsymbol{a}\omega, \quad \omega \equiv \|\boldsymbol{\omega}\|, \|\boldsymbol{a}\| = 1.$$

Due to the characteristic $\widehat{\boldsymbol{a}}^3 = -\widehat{\boldsymbol{a}}$, all $\widehat{\boldsymbol{a}}^n$ in the higher power terms can be replaced by $\widehat{\boldsymbol{a}}^2$. Furthermore, using the Taylor series of sin, cos we get the following equation.

$$e^{\widehat{\omega}t} = \boldsymbol{E} + \widehat{\boldsymbol{a}}\sin(\omega t) + \widehat{\boldsymbol{a}}^2(1 - \cos(\omega t)) \tag{2.36}$$

Equation (2.36) is called **Rodrigues' formula** and it gives you the rotation matrix directly from a constant angular velocity vector. This equation together with the basic equation of rotation are the most important equations in this book.

Equation (2.36) can be seen as giving us the rotation ωt [rad] around the axis defined by unit vector $\boldsymbol{a}$. By replacing the rotation angle using $\theta \equiv \omega t$, we obtain

$$e^{\widehat{a}\theta} = \boldsymbol{E} + \widehat{\boldsymbol{a}}\sin\theta + \widehat{\boldsymbol{a}}^2(1 - \cos\theta). \tag{2.37}$$

This equation is heavily used in kinematic calculation[7].

2.2.7 Matrix Logarithm

Having defined the matrix exponential, let us define it's inverse, the **matrix logarithm**

$$\ln \boldsymbol{R} = \ln e^{\widehat{\omega}} \equiv \widehat{\boldsymbol{\omega}}. \tag{2.38}$$

This is used to get the angular velocity vector from a given rotation matrix. The angular velocity vector integrated over one second will equate to the rotation described by the rotation matrix

$$\boldsymbol{\omega} = (\ln \boldsymbol{R})^\vee.$$

The actual calculation is

[7] Take note of its similarity with Euler's formula $e^{i\theta} = \cos\theta + i\sin\theta$, which was called **our jewel** by Richard Feynman [105].

$$(\ln \boldsymbol{R})^\vee = \begin{cases} [0\ 0\ 0]^T & (\text{if } \boldsymbol{R} = \boldsymbol{E}) \\ \frac{\pi}{2}\begin{bmatrix} r_{11}+1 \\ r_{22}+1 \\ r_{33}+1 \end{bmatrix} & (\text{else if } \boldsymbol{R} \text{ is diagonal}) \\ \theta \frac{\boldsymbol{l}}{\|\boldsymbol{l}\|} & (\text{otherwise}) \end{cases} \tag{2.39}$$

given

$$\boldsymbol{R} = \begin{bmatrix} r_{11}\ r_{12}\ r_{13} \\ r_{21}\ r_{22}\ r_{23} \\ r_{31}\ r_{32}\ r_{33} \end{bmatrix},$$

$$\boldsymbol{l} = \begin{bmatrix} r_{32} - r_{23} \\ r_{13} - r_{31} \\ r_{21} - r_{12} \end{bmatrix},$$

$$\theta = \text{atan2}(\|\boldsymbol{l}\|, r_{11} + r_{22} + r_{33} - 1)$$

where atan2(y, x) is the function that gives you the angle between vector (x, y) and the x axis. Please refer to the textbook [101] and the related paper [124] to know their exact derivation.

Using the matrix exponential and the matrix logarithm we can interpolate between two given rotation matrices $\boldsymbol{R}_1, \boldsymbol{R}_2$ as follows.

1. Get the rotation matrix which links the two matrices. $\boldsymbol{R} = \boldsymbol{R}_1^T \boldsymbol{R}_2$
2. Get the equivalent angular velocity vector from this rotation matrix. $\boldsymbol{\omega} = (\ln \boldsymbol{R})^\vee$
3. The angular velocity vector in world coordinates is: $\boldsymbol{R}_1\boldsymbol{\omega}$
4. The interpolation becomes $\boldsymbol{R}(t) = \boldsymbol{R}_1 e^{\widehat{\omega} t}, \quad t \in [0, 1]$.

2.3 Velocity in Three Dimensional Space

In this section we add translational motion to rotation and consider the velocity of an object moving freely in three dimensional space.

2.3.1 The Linear and Angular Velocity of a Single Object

A kinematic state of an object in 3D space can be described by the position of a reference point $\boldsymbol{p}$ in the object and a rotation matrix $\boldsymbol{R}$ for the attitude of the object as shown in Fig. 2.11. In other words, a pair of $(\boldsymbol{p}, \boldsymbol{R})$ represents a local coordinate system attached on the object. For a point on the object which locates at $\bar{\boldsymbol{p}}_k$ in the local coordinates, its position in the world coordinates is given by

$$\boldsymbol{p}_k = \boldsymbol{p} + \boldsymbol{R}\bar{\boldsymbol{p}}_k. \tag{2.40}$$

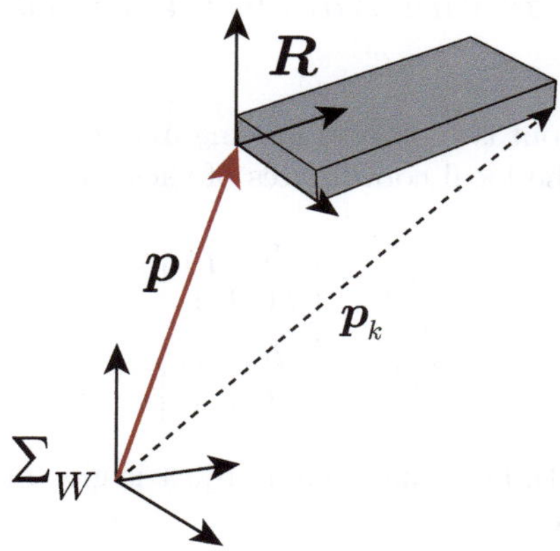

Fig. 2.11 Position and Attitude of an Object in 3D Space

Let us assume that this object is tumbling through space combining both translational and rotational motion. The velocity of the point $\boldsymbol{p}_k$ can be derived by differentiating the previous equation with respect to time

$$\begin{aligned}\dot{\boldsymbol{p}}_k &= \dot{\boldsymbol{p}} + \dot{\boldsymbol{R}}\bar{\boldsymbol{p}}_k \\ &= \boldsymbol{v} + \widehat{\boldsymbol{\omega}}\boldsymbol{R}\bar{\boldsymbol{p}}_k \\ &= \boldsymbol{v} + \boldsymbol{\omega} \times (\boldsymbol{R}\bar{\boldsymbol{p}}_k),\end{aligned} \tag{2.41}$$

where $\boldsymbol{v}, \boldsymbol{\omega}$ have been defined as follows

$$\boldsymbol{v} \equiv \dot{\boldsymbol{p}} \tag{2.42}$$

$$\boldsymbol{\omega} \equiv (\dot{\boldsymbol{R}}\boldsymbol{R}^T)^\vee. \tag{2.43}$$

By making substitutions in (2.41) using (2.40) we obtain

$$\dot{\boldsymbol{p}}_k = \boldsymbol{v} + \boldsymbol{\omega} \times (\boldsymbol{p}_k - \boldsymbol{p}). \tag{2.44}$$

The velocity of any point in the object can be calculated using this equation. Therefore we can conclude that:

> The motion of an object in 3D space can be represented by a 6 dimensional vector $[v_x\ v_y\ v_z\ \omega_x\ \omega_y\ \omega_z]^T$ which is a combination of its **linear velocity** $\boldsymbol{v}$ and **angular velocity** $\boldsymbol{\omega}$.

2.3.2 The Linear and Angular Velocity of Two Objects

Next we will think about two objects moving in 3D space. Let us assume that we have been given the local coordinates of these objects as follows

$$\boldsymbol{T}_1 = \begin{bmatrix} \boldsymbol{R}_1 & \boldsymbol{p}_1 \\ 0\,0\,0 & 1 \end{bmatrix} \tag{2.45}$$

$$^1\boldsymbol{T}_2 = \begin{bmatrix} \boldsymbol{R}_d & \boldsymbol{p}_d \\ 0\,0\,0 & 1 \end{bmatrix}. \tag{2.46}$$

The position and attitude of the second object is given to you relative to the first object as $^1\boldsymbol{T}_2$. So,

$$\begin{aligned} \boldsymbol{T}_2 &= \boldsymbol{T}_1\,^1\boldsymbol{T}_2 \\ &= \begin{bmatrix} \boldsymbol{R}_1 & \boldsymbol{p}_1 \\ 0\,0\,0 & 1 \end{bmatrix} \begin{bmatrix} \boldsymbol{R}_d & \boldsymbol{p}_d \\ 0\,0\,0 & 1 \end{bmatrix} \\ &= \begin{bmatrix} (\boldsymbol{R}_1\boldsymbol{R}_d) & (\boldsymbol{p}_1 + \boldsymbol{R}_1\boldsymbol{p}_d) \\ 0\,0\,0 & 1 \end{bmatrix}. \end{aligned} \tag{2.47}$$

Therefore the position and attitude of the second object in world coordinates is

$$\boldsymbol{p}_2 = \boldsymbol{p}_1 + \boldsymbol{R}_1\boldsymbol{p}_d \tag{2.48}$$

$$\boldsymbol{R}_2 = \boldsymbol{R}_1\boldsymbol{R}_d. \tag{2.49}$$

The linear velocity of the second object can be obtained by taking the time derivative of (2.48).

$$\begin{aligned} \boldsymbol{v}_2 &= \frac{d}{dt}(\boldsymbol{p}_1 + \boldsymbol{R}_1\boldsymbol{p}_d) \\ &= \dot{\boldsymbol{p}}_1 + \dot{\boldsymbol{R}}_1\boldsymbol{p}_d + \boldsymbol{R}_1\dot{\boldsymbol{p}}_d \\ &= \boldsymbol{v}_1 + \widehat{\boldsymbol{\omega}}_1\boldsymbol{R}_1\boldsymbol{p}_d + \boldsymbol{R}_1\boldsymbol{v}_d \\ &= \boldsymbol{v}_1 + \boldsymbol{\omega}_1 \times (\boldsymbol{R}_1\boldsymbol{p}_d) + \boldsymbol{R}_1\boldsymbol{v}_d \end{aligned}$$

where, $\boldsymbol{v}_1 \equiv \dot{\boldsymbol{p}}_1$, $\boldsymbol{v}_d \equiv \dot{\boldsymbol{p}}_d$.

Here, using (2.48) and rearranging its elements we obtain

$$\boldsymbol{v}_2 = \boldsymbol{v}_1 + \boldsymbol{R}_1\boldsymbol{v}_d + \boldsymbol{\omega}_1 \times (\boldsymbol{p}_2 - \boldsymbol{p}_1). \tag{2.50}$$

Let's take a moment to think about (2.22) in Section 2.2.4,

$$\boldsymbol{R}(\boldsymbol{\omega} \times \boldsymbol{p}) = (\boldsymbol{R}\boldsymbol{\omega}) \times (\boldsymbol{R}\boldsymbol{p}) \qquad \ldots (a)$$

The left side of this equation can be rewritten as

$$\begin{aligned} \boldsymbol{R}(\boldsymbol{\omega} \times \boldsymbol{p}) &= \boldsymbol{R}\widehat{\boldsymbol{\omega}}\boldsymbol{p} \\ &= \boldsymbol{R}\widehat{\boldsymbol{\omega}}\boldsymbol{R}^T\boldsymbol{R}\boldsymbol{p} \\ &= (\boldsymbol{R}\widehat{\boldsymbol{\omega}}\boldsymbol{R}^T)(\boldsymbol{R}\boldsymbol{p}). \qquad \ldots (b) \end{aligned}$$

Comparing equation (a) and equation (b) we can see that

$$(\boldsymbol{R}\boldsymbol{\omega})^\wedge = \boldsymbol{R}\widehat{\boldsymbol{\omega}}\boldsymbol{R}^T.$$

Fig. 2.12 Coordinate Transformation and the Angular Velocity Vector

The angular velocity of the second object can be calculated from (2.49) as follows

$$\begin{aligned} \widehat{\boldsymbol{\omega}}_2 &= \dot{\boldsymbol{R}}_2\boldsymbol{R}_2^T \\ &= \frac{d}{dt}(\boldsymbol{R}_1\boldsymbol{R}_d)\boldsymbol{R}_2^T \\ &= (\dot{\boldsymbol{R}}_1\boldsymbol{R}_d + \boldsymbol{R}_1\dot{\boldsymbol{R}}_d)\boldsymbol{R}_2^T \\ &= (\widehat{\boldsymbol{\omega}}_1\boldsymbol{R}_1\boldsymbol{R}_d + \boldsymbol{R}_1\widehat{\boldsymbol{\omega}}_d\boldsymbol{R}_d)\boldsymbol{R}_2^T \\ &= \widehat{\boldsymbol{\omega}}_1 + \boldsymbol{R}_1\widehat{\boldsymbol{\omega}}_d\boldsymbol{R}_d\boldsymbol{R}_d^T\boldsymbol{R}_1^T \\ &= \widehat{\boldsymbol{\omega}}_1 + \boldsymbol{R}_1\widehat{\boldsymbol{\omega}}_d\boldsymbol{R}_1^T \end{aligned}$$

where $\boldsymbol{\omega}_1 \equiv (\dot{\boldsymbol{R}}_1\boldsymbol{R}_1^T)^\vee$, $\boldsymbol{\omega}_d \equiv (\dot{\boldsymbol{R}}_d\boldsymbol{R}_d^T)^\vee$.

From Fig. 2.12 we can see that $(\boldsymbol{R}\boldsymbol{\omega})^\wedge = \boldsymbol{R}\widehat{\boldsymbol{\omega}}\boldsymbol{R}^T$, so we get

$$\widehat{\boldsymbol{\omega}}_2 = \widehat{\boldsymbol{\omega}}_1 + (\boldsymbol{R}_1\boldsymbol{\omega}_d)^\wedge.$$

By applying $\vee$ to both sides of this equation we get,

$$\boldsymbol{\omega}_2 = \boldsymbol{\omega}_1 + \boldsymbol{R}_1\boldsymbol{\omega}_d. \qquad (2.51)$$

In summary, we first defined the position and attitude of object 1 and object 2 as $(\boldsymbol{p}_1, \boldsymbol{R}_1)$, $(\boldsymbol{p}_2, \boldsymbol{R}_2)$. If the velocity of object 1 is $(\boldsymbol{v}_1, \boldsymbol{\omega}_1)$ and the velocity of object 2 relative to object 1 is $(\boldsymbol{v}_d, \boldsymbol{\omega}_d)$, the velocity of object 2 will be given as follows[8]

[8] The linear and angular velocity of the object have simply been combined and called "velocity".

$$\boldsymbol{v}_2 = \boldsymbol{v}_1 + \boldsymbol{R}_1 \boldsymbol{v}_d + \boldsymbol{\omega}_1 \times (\boldsymbol{p}_2 - \boldsymbol{p}_1) \tag{2.52}$$
$$\boldsymbol{\omega}_2 = \boldsymbol{\omega}_1 + \boldsymbol{R}_1 \boldsymbol{\omega}_d. \tag{2.53}$$

Here $\boldsymbol{R}_1\boldsymbol{v}_d$ and $\boldsymbol{R}_1\boldsymbol{\omega}_d$ describe the relative velocity between the objects in world coordinates. If we replace these using ${}^W\boldsymbol{v}_d, {}^W\boldsymbol{\omega}_d$ we get,

$$\boldsymbol{v}_2 = \boldsymbol{v}_1 + {}^W\boldsymbol{v}_d + \boldsymbol{\omega}_1 \times (\boldsymbol{p}_2 - \boldsymbol{p}_1) \tag{2.54}$$
$$\boldsymbol{\omega}_2 = \boldsymbol{\omega}_1 + {}^W\boldsymbol{\omega}_d. \tag{2.55}$$

Therefore barring the fact that we get an influence of the rotational velocity the third element of (2.54), the velocity of the object in world coordinates is simply a sum.

In Fig. 2.13 we have tried to visualise the meaning of (2.55).

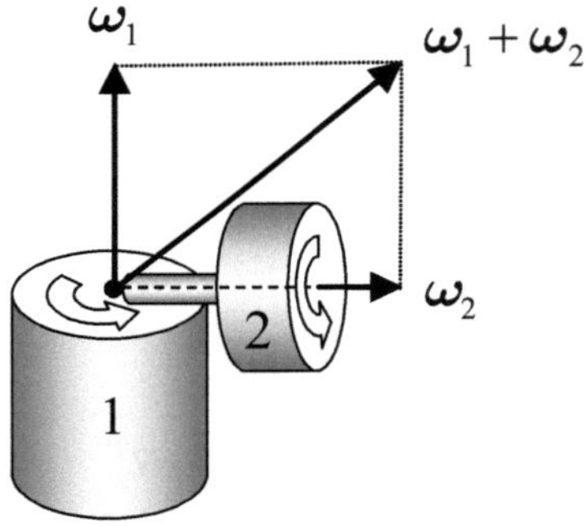

Fig. 2.13 The sum of angular velocity vectors: Object 1 rotates according to angular velocity vector $\boldsymbol{\omega}_1$. Object 2 rotates relative to Object 1 at $\boldsymbol{\omega}_2$. The angular velocity of Object 2 in world coordinates is $\boldsymbol{\omega}_1 + \boldsymbol{\omega}_2$.

2.4 Robot Data Structure and Programming

2.4.1 Data Structure

A humanoid robot is a mechanism consisting of many links connect by joints. To start our analyze, we would like to divide it into smaller units. Among many possible ways of separation, we present two typical ways in Fig. 2.14.

In Fig. 2.14(a) each joint gets included in the link that is further away from the trunk. In (b) each joint is defined as a part of the trunk or the link closer to the trunk. In terms of computer programming the former style is more convenient. The reason is because all links will include just one joint so you can use the same algorithm to deal with all the joints. In addition it also makes adding new joints easier. On the other hand, method (b) will

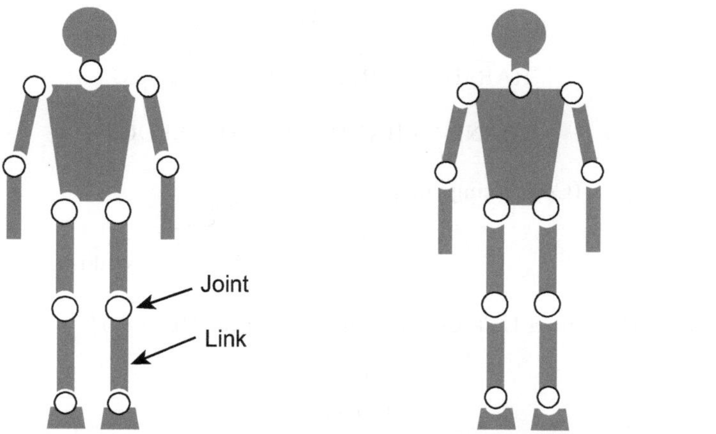

Fig. 2.14 Dividing a robot structure. In (a) all links have one joint except for the trunk. In (b) the number of joints in each link differs with each joint. From a programming perspective the former is easier to deal with.

require links with different numbers of joints. Which in turn makes the computer programming more complex. Please notice that here we are discussing a *conceptual separation* of a robot. It is not an absolute requirement to divide the robot as cleanly as shown in Fig. 2.14.

By joining these links in the correct combination we get the composition of links that forms a humanoid robot. The connection rule is presented by a diagram shown in Fig. 2.15 which has a form a **tree structure**[9]. In other words, we could probably say that this structure shows the family tree of links with the base link as a common ancestor with the extremities being the youngest members.

However, in the graph of Fig. 2.15 there exist different numbers of branches (connections) depending on the link, so as we discussed regarding dividing a robot structure, this representation makes programming difficult.

Alternatively, we could describe the link connection as shown in Fig. 2.16. In this graph, each link has two connections so that the left lower line connects to its child link and the right lower line connects to its sister link[10]. For example, if you want to know the sister links of RARM, you go along the lower right connections and gain the links, "LARM", "RLEG" and "LLEG". The parent of these links will be the upper right "BODY" link and the child will be the lower left "RHAND" link. At the ends which have no child or sister link we have used "0". In conclusion, although this graph looks different, it represents the same information represented by the graph in Fig. 2.15.

[9] A tree graph is a graph with the structure that it has a base from which smaller branches branch out without forming loops.

[10] Usually the word "sibling" is used for this purpose.

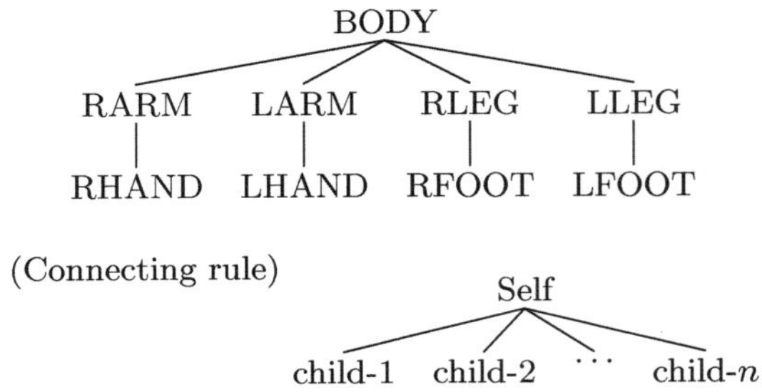

Fig. 2.15 Humanoid Link Connection Description Method 1[11]

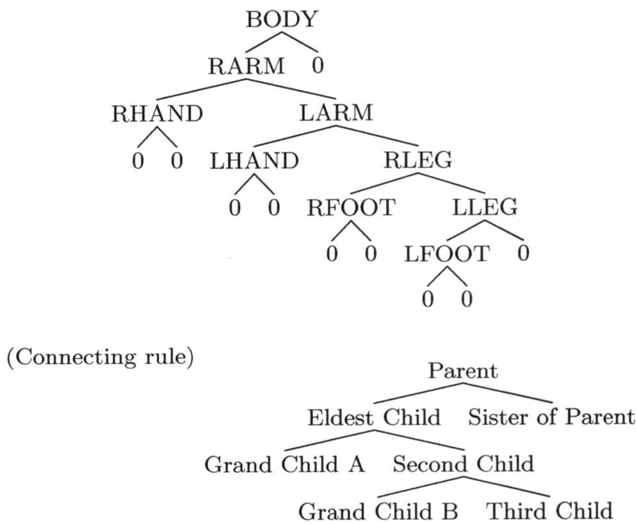

Fig. 2.16 Humanoid Link Connection Description Method 2

2.4.2 Programming with Recursions

Next let us look at how we would actually realise a program which would use these data structures. In preparation, we will convert the information in Fig. 2.16 into a list. Starting with ID number 1 which we use for the base of this tree, we have given each link an ID number and used them in the following list.

[11] When naming the links in this textbook, we use *R* or *L* as a prefix to show which side the link is on. So the links in the right arm will be *RARM* and the left arm will be *LARM*.

ID	name	sister	child
1	BODY	0	2
2	RARM	4	3
3	RHAND	0	0
4	LARM	6	5
5	LHAND	0	0
6	RLEG	8	7
7	RFOOT	0	0
8	LLEG	0	9
9	LFOOT	0	0

To input this information in Matlab we should type as following.[12]

```
>> global uLINK
>> uLINK(1).name = 'BODY';
>> uLINK(1).sister = 0;
>> uLINK(1).child = 2;
```

The ">>" above is the Matlab command prompt. uLINK is the name of a structure used in this textbook, which holds most of the values required in the calculations. Here we only look at the fields, "name", "sister" and "child". In the first line we declare that we want to use the globally available variable uLINK. In the next lines, uLINK(1) points to the first uLINK with the ID number 1. The names after the period specify data field names (name, sister, child).

In the same way you access the fields in the part which holds the values link with ID number 2 though uLINK(2).[13]

```
>> uLINK(1).name          % print the name of the link with ID=1
ans =
BODY
>> uLINK(uLINK(1).child).name   % print first child of BODY
ans =
RARM
>> uLINK(uLINK(uLINK(1).child).child).name % first grandchild of BODY
ans =
RHAND
```

[12] In this textbook, we will use Matlab, a product of MathWorks which enables the user to do matrix calculations interactively. The algorithm itself does not depend on the language you use, so this program can also be easily implemented in other languages such as C++ or Java.

[13] Once the program starts to get longer you would store these commands in a file with a suffix "m", and execute this from the command line.

If you omit the semi-colon ";" after the name of a variable on the command line, Matlab will print out all information regarding that variable. By using this you can check the information stored in this tree whenever you want. In other words you now have a database of the robot link information.

Figure 2.17 shows you a short script (PrintLinkName.m) which prints all the names of the links stored in the database.

```
function PrintLinkName(j)
global uLINK  % Refer global variable uLINK

if j ~= 0
    fprintf('j=%d : %s\n',j,uLINK(j).name); % Show my name
    PrintLinkName(uLINK(j).child);        % Show my child's name
    PrintLinkName(uLINK(j).sister);       % Show my sister's name
end
```

Fig. 2.17 PrintLinkName.m: Show the name of all links

This program takes an ID number as input, and check it first. If the ID is not 0, the program prints the corresponding link name and calls PrintLinkName with the sister ID number, and then calls PrintLinkName using the child ID number. If the ID is 0, the program does nothing. After saving this script with the name PrintLinkName.m, we can execute "PrintLinkName(1)" on the Matlab command line and get the names of the child links within the tree under the "BODY" link. Calling a function from inside it is named a **recursive call**. Each time this function is called you move downward in the tree until you reach a node which has neither sister nor child. This gives you a simple way to visit each node in the tree and to apply a desired computation.

Using the technique of the recursive call, we can easily write a function to sum up all masses of the links (Fig. 2.18). Here we assume a new data field "m" was added to hold the mass of the link and it was initialized with the actual mass of each link.

If the ID number is 0, then the program will return 0 kg because there are no links to sum (line 4). If the ID number is not 0, then it will return the sum of the mass of the corresponding link, the total mass of the descendant links from the sister, and the total mass of the descendant links from the child (line 6). If you type TotalMass(1) on the command line, this function will visit all nodes in the tree and return the total mass of the links under the "BODY" link. Compared with the usual method which checks the existing links beforehand and sums them, I feel this is much smarter. What do you think?

```
function m = TotalMass(j)
global uLINK

if j == 0
   m = 0;
else
   m = uLINK(j).m + TotalMass(uLINK(j).sister) + TotalMass(uLINK(j).child);
end
```

Fig. 2.18 TotalMass.m: Sum of Each Link's Mass

2.5 Kinematics of a Humanoid Robot

2.5.1 Creating the Model

To explain the kinematics of a humanoid robot, we will use a 12 DoF model shown in Fig. 2.19, which consists of two legs. The link names and their ID numbers are indicated in Fig. 2.19(a). Splitting this model by the manner of Fig. 2.14(a), each link will have just one joint to drive it except the BODY link. Thus we can identify a link by joint name or its ID number. For example,

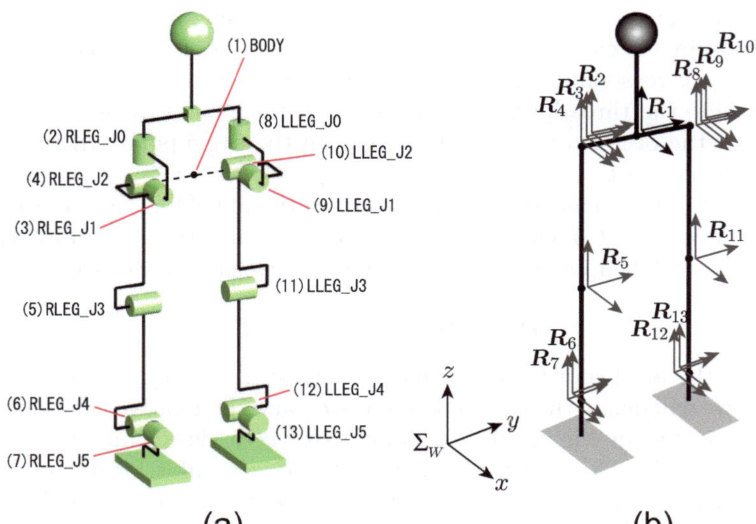

Fig. 2.19 (a) Structure of a 12 degree of freedom biped robot. Numbers in brackets refer to the ID number (b) Rotation matrix which describes attitude of each link.

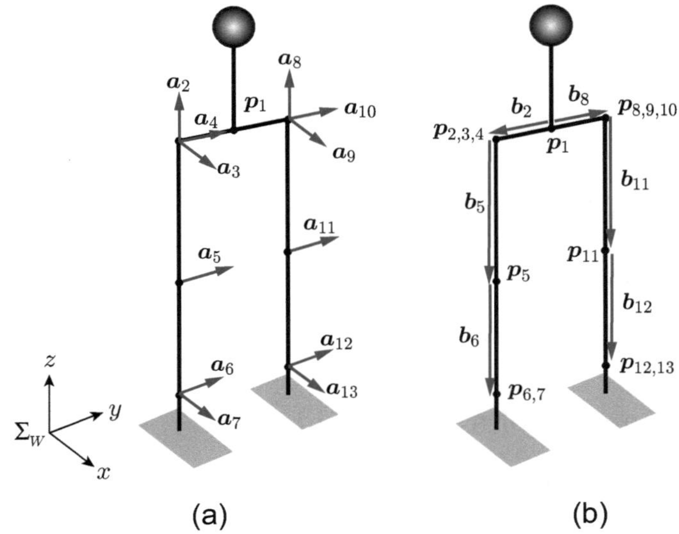

Fig. 2.20 (a) Joint Axis Vector $\boldsymbol{a}_j$ (b) Relative Position Vector $\boldsymbol{b}_j$ and the Origin of Local Coordinates $\boldsymbol{p}_j$

by the ID number 5, we refer the joint RLEG_J3 as well as the link of the right lower leg.

First, we must define the local coordinates for each link. An origin of each local coordinates can be set anywhere on its own joint axis. For each hip, however, it is reasonable to assign the origins of the three frames at the same point where the three joint axes intersect. In the same way, for each ankle, we assign the origins of two ankle frames on the same point where the two joint axes intersect.

Then, all rotation matrices which describe attitude of links are set to match the world coordinates when the robot is in its initial state, standing upright with fully stretched knees. Therefore we set thirteen matrices as,

$$\boldsymbol{R}_1 = \boldsymbol{R}_2 = \ldots = \boldsymbol{R}_{13} = \boldsymbol{E}.$$

We show the local coordinates defined at this stage in Fig. 2.19(b).

Next we will define the joint axis vectors $\boldsymbol{a}_j$ and the relative position vector $\boldsymbol{b}_j$ as indicated in Fig. 2.20. The joint axis vector is a unit vector which defines the axis of the joint rotation in the parent link's local coordinates. The positive (+) joint rotation is defined as the way to tighten a right-hand screw placed in the same direction with the joint axis vector. Using the knee joints as an example, the joint axis vectors would be, $\boldsymbol{a}_5, \boldsymbol{a}_{11} = [0\ 1\ 0]^T$. When we rotate this link in the + direction, a straight knee will flex in the same direction as a human's. The relative position vector $\boldsymbol{b}_j$ is the vector that indicates where the origin of a local coordinate lies in the parent link's local coordinates. When they lie in the same place as it is in the case of the ankle roll joint, $\boldsymbol{b}_7, \boldsymbol{b}_{13} = \boldsymbol{0}$[14].

In the following section we will use the description above to calculate Forward Kinematics, the Jacobian Matrix and Inverse Kinematics. To do this we will need a lot more information such as the shape, joint angle, joint velocity, etc. The full list of link parameters is shown in Table 2.1.

Table 2.1 Link Parameters

Link Parameter	Symbol for Equation	uLINK data field
Self ID	j	-
Sister ID	None	sister
Child ID	None	child
Parent ID	i	mother
Position in World Coordinates	$\boldsymbol{p}_j$	p
Attitude in World Coordinates	$\boldsymbol{R}_j$	R
Linear Velocity in World Coordinates	$\boldsymbol{v}_j$	v
Angular Velocity in World Coordinates	$\boldsymbol{\omega}_j$	w
Joint Angle	q_j	q
Joint Velocity	$\dot{q}_j$	dq
Joint Acceleration	$\ddot{q}_j$	ddq
Joint Axis Vector(Relative to Parent)	$\boldsymbol{a}_j$	a
Joint Relative Position(Relative to Parent)	$\boldsymbol{b}_j$	b
Shape(Vertex Information, Link Local)	$\bar{\boldsymbol{p}}_j$	vertex
Shape(Vertice Information (Point Connections)	None	face
Mass	m_j	m
Center of Mass(Link Local)	$\bar{\boldsymbol{c}}_j$	c
Moment of Inertia(Link Local)	$\bar{\boldsymbol{I}}_j$	I

2.5.2 Forward Kinematics: Calculating the Position of the Links from Joint Angles

Forward Kinematics is a calculation to obtain the position and attitude of a certain link from a given robot structure and its joint angles. This is required when you want to calculate the Center of Mass of the whole robot, when you just want to display the current state of the robot, or when you want to detect collisions of the robot with the environment. Thus forward kinematics forms the basis of robotics simulation.

Forward kinematics can be calculated by using the chain rule of homogeneous transforms. First we will start off by calculating the homogeneous

[14] There is a well known method you can use to describe the link structure of a robot called the Denavit-Hartenberg (DH) method [18]. We also used this method at first. However this method has a restriction which requires you to change the orientation of the link coordinates with each link. We found the implementation with this restriction to be rather error-prone, so we instead adopted the method outlined above.

transform of a single link as shown in Fig. 2.21. We need to set a local coordinate system Σ_j which has it's origin on the joint axis. The joint axis vector seen from the parent coordinates is $\boldsymbol{a}_j$ and the origin of Σ_j is $\boldsymbol{b}_j$. The joint angle is q_j and the attitude of the link when the joint angle is 0 is, $\boldsymbol{E}$.

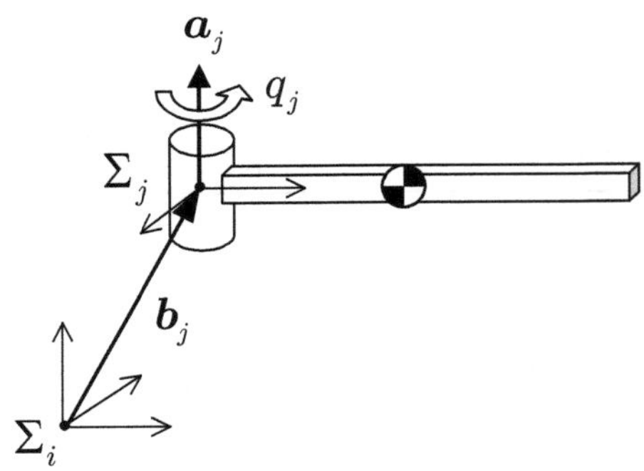

Fig. 2.21 The Position, Attitude and Rotation of a single link $\boldsymbol{a}_j, \boldsymbol{b}_j$ each specify the joint axis vector and location of the origin viewed from the parent coordinate system

The homogeneous transform relative to the parent link is:

$$ {}^i\boldsymbol{T}_j = \begin{bmatrix} e^{\widehat{a}_j q_j} & \boldsymbol{b}_j \\ 0\,0\,0 & 1 \end{bmatrix}. \tag{2.56}$$

Next let us assume there are two links as shown in Fig. 2.22. We will assume that the absolute position and attitude of the parent link $\boldsymbol{p}_i$, $\boldsymbol{R}_i$ is known. Therefore, the homogeneous transform to Σ_i becomes:

$$\boldsymbol{T}_i = \begin{bmatrix} \boldsymbol{R}_i & \boldsymbol{p}_i \\ 0\,0\,0 & 1 \end{bmatrix}. \tag{2.57}$$

From the chain rule of homogeneous transforms Σ_j is:

$$\boldsymbol{T}_j = \boldsymbol{T}_i {}^i\boldsymbol{T}_j. \tag{2.58}$$

From (2.56), (2.57) and (2.58) the absolute position ($\boldsymbol{p}_j$) and attitude ($\boldsymbol{R}_j$) of Σ_j can be calculated as being,

$$\boldsymbol{p}_j = \boldsymbol{p}_i + \boldsymbol{R}_i \boldsymbol{b}_j \tag{2.59}$$

$$\boldsymbol{R}_j = \boldsymbol{R}_i e^{\widehat{a}_j q_j} \tag{2.60}$$

Using this relationship and the recursive algorithm, the Forward Kinematics can be performed by an extremely simple script shown in Fig. 2.23. To use this program, we first set the absolute position and attitude of the base

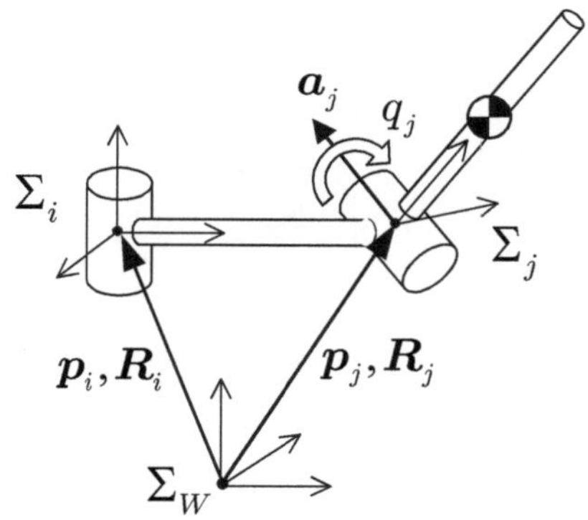

Fig. 2.22 Relative Position of Two Links

```
function ForwardKinematics(j)
global uLINK

if j == 0 return; end
if j ~= 1
    i = uLINK(j).mother;
    uLINK(j).p = uLINK(i).R * uLINK(j).b + uLINK(i).p;
    uLINK(j).R = uLINK(i).R * Rodrigues(uLINK(j).a, uLINK(j).q);
end
ForwardKinematics(uLINK(j).sister);
ForwardKinematics(uLINK(j).child);
```

Fig. 2.23 ForwardKinematics.m calculate forward kinematics for all joints

link (BODY) and all joint angles. Then by executing ForwardKinematics(1), we can update the positions and attitudes of all links in the robot.

Figure 2.24 shows what the 12 degree of freedom biped robot looks like when you give it random values for joint angles to all 12 joints. This should help you to imagine how a simple mechanism embodies a large amount of complexity.

2.5.3 Inverse Kinematics: Calculating the Joint Angles from a Link's Position and Attitude

Next we will discuss how to calculate joint angles when we have the position and attitude of the body and the foot to realize. What we need to do in this cases is **Inverse Kinematics**. For instance, suppose our robot is in the front of stairs and we want to place one of the foot on the first step whose height and depth are already known. We certainly need to determine the amount of

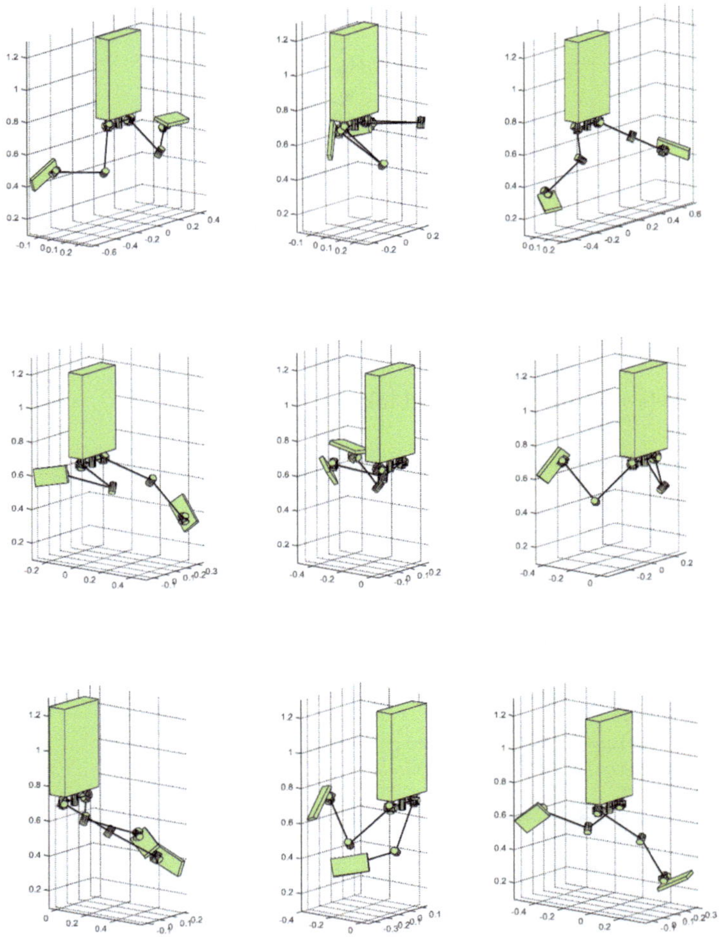

Fig. 2.24 Random poses calculated using ForwardKimematics. The knees are limited to $[0,\ \pi]$[rad], other joints are restricted to $[-\frac{\pi}{3},\ \frac{\pi}{3}]$[rad].

joint rotation for the hip, knee and the ankle. Inverse Kinematics is necessary for such a case.

There exist both an analytical method and a numerical method of solving Inverse Kinematics. First we will explain how to solve it analytically. Let's focus on the right leg of the model shown in Fig. 2.19. The position and attitude of the body and right leg will be $(\boldsymbol{p}_1, \boldsymbol{R}_1)$ and $(\boldsymbol{p}_7, \boldsymbol{R}_7)$ respectively. To simplify the equation we will define D, which is the distance between the Body origin and the hip joint. The upper leg length is A, and the lower leg length is B, as shown in Fig. 2.25(a). So therefore, the position of the hip would be

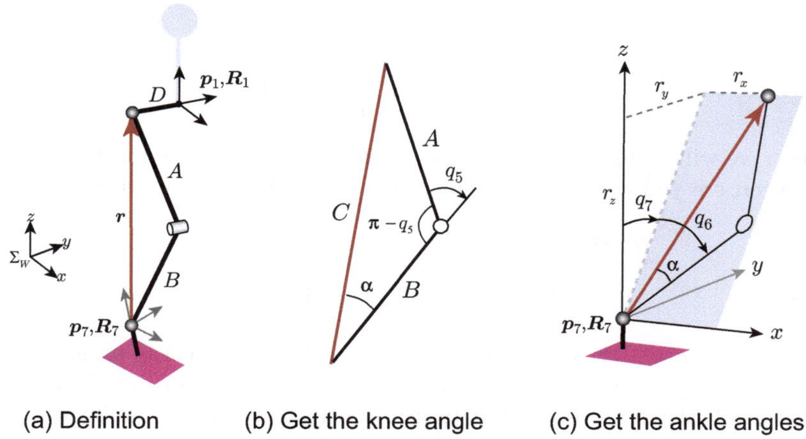

Fig. 2.25 Calculating Inverse Kinematics of the Legs

$$p_2 = p_1 + R_1 \begin{bmatrix} 0 \\ D \\ 0 \end{bmatrix}.$$

Next, we calculate the position of the crotch viewed from the ankle coordinate space

$$\boldsymbol{r} = \boldsymbol{R}_7^T(\boldsymbol{p}_2 - \boldsymbol{p}_7) \equiv [r_x \; r_y \; r_z]^T. \tag{2.61}$$

From this we can calculate the distance between the ankle and the hip, which we will define as

$$C = \sqrt{r_x^2 + r_y^2 + r_z^2}.$$

As shown in Fig. 2.25(b), if we consider the triangle ABC we get the angle of the knee q_5. From the **cosine rule** we get,

$$C^2 = A^2 + B^2 - 2AB\cos(\pi - q_5).$$

So the angle of the knees will be,

$$q_5 = -\cos^{-1}\left(\frac{A^2 + B^2 - C^2}{2AB}\right) + \pi.$$

If we define the angle at the lower end of the triangle as α, from the **sine rule** we get,

$$\frac{C}{\sin(\pi - q_5)} = \frac{A}{\sin\alpha}.$$

So therefore,

$$\alpha = \sin^{-1}\left(\frac{A\sin(\pi - q_5)}{C}\right).$$

Next we will focus on the ankle local coordinates. As shown in Fig. 2.25(c), from vector $\boldsymbol{r}$ you can calculate the ankle roll and pitch angles. So,

$$\begin{aligned} q_7 &= \text{atan2}(r_y, r_z) \\ q_6 &= -\text{atan2}\left(r_x, \text{sign}(r_z)\sqrt{r_y^2 + r_z^2}\right) - \alpha. \end{aligned}$$

The function atan2(y, x) calculates the angle between vector (x, y) and the x axis as it has already appeared in Section 2.2.7. It is built into Matlab and is also available as a built-in function in most programming languages. Also $sign(x)$ is a function that returns $+1$ if x is a positive value and -1 if it is negative.

What remains is the yaw, roll and pitch angles at the base of the leg. From the equations that define each joint

$$\boldsymbol{R}_7 = \boldsymbol{R}_1 \boldsymbol{R}_z(q_2)\boldsymbol{R}_x(q_3)\boldsymbol{R}_y(q_4)\boldsymbol{R}_y(q_5 + q_6)\boldsymbol{R}_x(q_7),$$

we obtain

$$\boldsymbol{R}_z(q_2)\boldsymbol{R}_x(q_3)\boldsymbol{R}_y(q_4) = \boldsymbol{R}_1^T \boldsymbol{R}_7 \boldsymbol{R}_x(-q_7)\boldsymbol{R}_y(-q_5 - q_6).$$

By expanding the left side of this equation and calculating the right hand side we get the following

$$\begin{bmatrix} c_2c_4 - s_2s_3s_4 & -s_2c_3 & c_2s_4 + s_2s_3c_4 \\ s_2c_4 + c_2s_3s_4 & c_2c_3 & s_2s_4 - c_2s_3c_4 \\ -c_3s_4 & s_3 & c_3c_4 \end{bmatrix} = \begin{bmatrix} R_{11} & R_{12} & R_{13} \\ R_{21} & R_{22} & R_{23} \\ R_{31} & R_{32} & R_{33} \end{bmatrix}$$

where $c_2 \equiv \cos q_2$, and $s_2 \equiv \sin q_2$.

By looking carefully at the left hand side of this equation we get,

$$q_2 = \text{atan2}(-R_{12}, R_{22}) \tag{2.62}$$

$$q_3 = \text{atan2}(R_{32}, -R_{12}s_2 + R_{22}c_2) \tag{2.63}$$

$$q_4 = \text{atan2}(-R_{31}, R_{33}). \tag{2.64}$$

An implementation of the above is shown in Fig. 2.26[15]. For the left leg, we could invert the sign on D and apply the same program.

This implementation can only be applied to a robot which has the same layout as the one in Fig. 2.19. If the robot does not have three joint axes that intersect each other at one point, we need an entirely different algorithm. There are many different algorithms outlined in the robot textbooks

[15] As a practical implementation, our program covers the target position exceeding the leg length and joint angle limits.

```
function q = IK_leg(Body,D,A,B,Foot)

r = Foot.R' * (Body.p + Body.R * [0 D 0]'- Foot.p);  % crotch from ankle
C = norm(r);
c5 = (C^2-A^2-B^2)/(2.0*A*B);
if c5 >= 1
    q5 = 0.0;
elseif c5 <= -1
    q5 = pi;
else
    q5 = acos(c5);  % knee pitch
end
q6a = asin((A/C)*sin(pi-q5));    % ankle pitch sub

q7 = atan2(r(2),r(3));  % ankle roll -pi/2 < q(6) < pi/2
if q7 > pi/2, q7=q7-pi; elseif q7 < -pi/2, q7=q7+pi; end
q6 = -atan2(r(1),sign(r(3))*sqrt(r(2)^2+r(3)^2)) -q6a; % ankle pitch

R = Body.R' * Foot.R * Rroll(-q7) * Rpitch(-q6-q5); %% hipZ*hipX*hipY
q2  = atan2(-R(1,2),R(2,2));    % hip yaw
cz = cos(q2); sz = sin(q2);
q3 = atan2(R(3,2),-R(1,2)*sz + R(2,2)*cz);  % hip roll
q4 = atan2( -R(3,1), R(3,3));               % hip pitch

q = [q2 q3 q4 q5 q6 q7]';
```

Fig. 2.26 IK_leg.m Example implementation of an analytical solution to Inverse Kinematics. CAUTION! When using this program on a real robot, you need to continuously check whether the joint angles exceed their limits. In the worst case, you could destroy your robot or otherwise cause major injury or death.

(for instance [96, 127]), but in general it requires a large amount of heavy calculation, so it is more common to use the numerical solution which we will go over in the next section.

2.5.4 Numerical Solution to Inverse Kinematics

Compared to solving the inverse kinematics analytically, forward kinematics calculation is simple (Section 2.5.2). So it isn't all that far fetched to think of a trial and error method of solving the inverse kinematics by using forward kinematics, as shown in Fig. 2.27. A sample algorithm could be,

Step 1. Prepare the position and attitude $(\boldsymbol{p}^{ref}, \boldsymbol{R}^{ref})$ of the base link
Step 2. Prepare the position and attitude $(\boldsymbol{p}^{ref}, \boldsymbol{R}^{ref})$ of the target link

Step 3. Define vector $\boldsymbol{q}$ which holds the joint angles from the base link to the target link
Step 4. Use forward kinematics to calculate the position and attitude $(\boldsymbol{p}, \boldsymbol{R})$ of the target link
Step 5. Calculate the difference in position and attitude $(\Delta \boldsymbol{p}, \Delta \boldsymbol{R}) = (\boldsymbol{p}^{ref} - \boldsymbol{p}, \boldsymbol{R}^T \boldsymbol{R}^{ref})$
Step 6. If $(\Delta \boldsymbol{p}, \Delta \boldsymbol{R})$ are small enough stop the calculation
Step 7. If $(\Delta \boldsymbol{p}, \Delta \boldsymbol{R})$ are not small enough calculate $\Delta \boldsymbol{q}$ which would reduce the error
Step 8. Update joint angles by $\boldsymbol{q} := \boldsymbol{q} + \Delta \boldsymbol{q}$ and return to **Step 4**

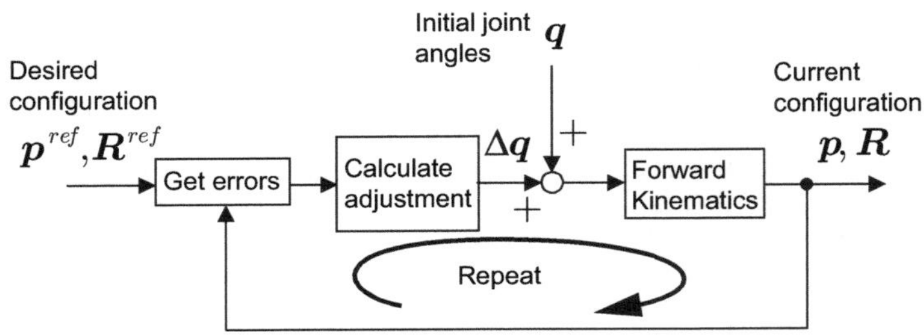

Fig. 2.27 Basic Concept Behind Numerical Approach to Inverse Kinematics: Use forward kinematics to and adjust the joint angles to narrow the difference.

To actually implement this you first need to surmount the next two hurdles.

1. What do we really mean by the position and attitude errors $(\Delta \boldsymbol{p}, \Delta \boldsymbol{R})$ being small enough? (**Step 6**)
2. How do we actually go about calculating $\Delta \boldsymbol{q}$, to narrow the gap? (**Step 7**)

The first problem can be solved relatively easily. Zero position error and zero attitude error can be described with the following equations

$$\Delta \boldsymbol{p} = 0$$
$$\Delta \boldsymbol{R} = \boldsymbol{E}.$$

An example function which returns positive scalar depending on the error magnitude is the following[16]

[16] A more general function would be $err(\Delta \boldsymbol{p}, \Delta \boldsymbol{R}) = \alpha \|\Delta \boldsymbol{p}\|^2 + \beta \|\Delta \boldsymbol{\theta}\|^2$. Here α and β are some positive number with the direction requiring more precision being larger.

$$err(\Delta \boldsymbol{p}, \Delta \boldsymbol{R}) = \|\Delta \boldsymbol{p}\|^2 + \|\Delta \boldsymbol{\theta}\|^2, \tag{2.65}$$
$$\Delta \boldsymbol{\theta} \equiv (\ln \Delta \boldsymbol{R})^{\vee}. \tag{2.66}$$

This function becomes zero only at both position and attitude error is zero. You can say the position and attitude errors are small enough when $err(\Delta \boldsymbol{p}, \Delta \boldsymbol{R})$ becomes smaller than predefined value, for example 1×10^{-6}.

How about the second problem? All we really need to do is come up with a set of joint angles $\Delta \boldsymbol{q}$ which lowers $err(\Delta \boldsymbol{p}, \Delta \boldsymbol{R})$. One idea would be to use random numbers each time. If we are able to lower $err(\Delta \boldsymbol{p}, \Delta \boldsymbol{R})$ by even a small amount, we will use it for the joint angles and start over. The robot uses trial and error to search for the joint angles itself so we have the **illusion of intelligence**[17].

Although it is an enticing idea, none of the robots today would use this method. The reason is that there is a method which is much faster and far more precise. In this method which is called the Newton-Raphson method we first start off by considering what happens to the position and attitude $(\delta \boldsymbol{p}, \delta \boldsymbol{\theta})$ when you change the joint angles using a minute value of $\delta \boldsymbol{q}$

$$\delta \boldsymbol{p} = \boldsymbol{X}_p(\boldsymbol{q}, \delta \boldsymbol{q}) \tag{2.67}$$
$$\delta \boldsymbol{\theta} = \boldsymbol{X}_\theta(\boldsymbol{q}, \delta \boldsymbol{q}). \tag{2.68}$$

Here, $\boldsymbol{X}_p$ and $\boldsymbol{X}_\theta$ are unknown, but when $\delta \boldsymbol{q}$ is small let us say that we can describe it simply with addition and multiplication. If we use a matrix we get,

$$\begin{bmatrix} \delta \boldsymbol{p} \\ \delta \boldsymbol{\theta} \end{bmatrix} = \begin{bmatrix} J_{11} & J_{12} & J_{13} & J_{14} & J_{15} & J_{16} \\ J_{21} & J_{22} & J_{23} & J_{24} & J_{25} & J_{26} \\ J_{31} & J_{32} & J_{33} & J_{34} & J_{35} & J_{36} \\ J_{41} & J_{42} & J_{43} & J_{44} & J_{45} & J_{46} \\ J_{51} & J_{52} & J_{53} & J_{54} & J_{55} & J_{56} \\ J_{61} & J_{62} & J_{63} & J_{64} & J_{65} & J_{66} \end{bmatrix} \delta \boldsymbol{q}. \tag{2.69}$$

Here $J_{ij}, (i, j = 1 \ldots 6)$ are constants which are defined by the current position and attitude of the robots links. There are 6 because of the number of links in the leg. It is too much to write all the components each time so we will simplify it by

$$\begin{bmatrix} \delta \boldsymbol{p} \\ \delta \boldsymbol{\theta} \end{bmatrix} = \boldsymbol{J}\, \delta \boldsymbol{q}. \tag{2.70}$$

[17] This idea is something anyone would think of, but for some reason a lot of people tend to think that they are the only ones to think of it. Actually the author happened to be one of them :-).

The matrix $\boldsymbol{J}$ is called the **Jacobian**[18]. Once we have (2.70) we can calculate the required adjustment by simply taking the inverse of this matrix

$$\delta\boldsymbol{q} = \lambda\, \boldsymbol{J}^{-1} \begin{bmatrix} \delta\boldsymbol{p} \\ \delta\boldsymbol{\theta} \end{bmatrix}. \tag{2.71}$$

This is the equation to calculate the adjustments of the joint angles based on the errors in position and attitude. The value $\lambda \in (0\ 1]$ is a coefficient used to stabilize the numeric calculation. Figure 2.28 shows a sample implementation of the inverse kinematics algorithm written in Matlab. On the 7th line you will see the function CalcJacobian which is used to calculate the Jacobian. We will go over this in more detail in the next section. The 10th line is the actual implementation of (2.71). The operator \ "backslash" efficiently solves the linear equations without doing explicit matrix inversion.

```
function InverseKinematics(to, Target)
global uLINK

lambda = 0.5;
ForwardKinematics(1);
idx = FindRoute(to);
for n = 1:10
  J   = CalcJacobian(idx);
  err = CalcVWerr(Target, uLINK(to));
  if norm(err) < 1E-6 return, end;
  dq = lambda * (J \ err);
  for nn=1:length(idx)
    j = idx(nn);
    uLINK(j).q = uLINK(j).q + dq(nn);
  end
  ForwardKinematics(1);
end
```

Fig. 2.28 InvserseKinematics.m Numerical Solution to Inverse Kinematics

The function FindRoute returns the links that you need to go through to get to the target link from the base link. CalcVWerr is a function which calculates the difference in position and attitude. You can find implemented versions of these functions in the appendix at the end of this chapter.

[18] From the German mathematician Carl Gustav Jacobi (1804-1851). When mathematicians refer to the Jacobian it means the determinant of this matrix, but when roboticists talk about the Jacobian they usually mean the matrix itself. Some people think that this is a mistake but this is not something that is local to Japan, it is done the world over.

We show an example use of Inverse Kinematics as Matlab command input in Fig. 2.29. Here we use `SetupBipedRobot` to set robot data, `GoHalfSitting` to get non-singular posture, and `DrawAllJoints()` to display the biped robot. The function `rpy2rot()` is a implementation of (2.13). They can be obtained from the download material.

```
>> SetupBipedRobot;   % Set robot parameters
>> GoHalfSitting;     % Set knee bending posture

>> Rfoot.p = [-0.3, -0.1, 0]';
>> Rfoot.R = rpy2rot(0, ToRad*20.0,0);
>> InverseKinematics(RLEG_J5, Rfoot);

>> Lfoot.p = [ 0.3, 0.1, 0]';
>> Lfoot.R = rpy2rot(0, -ToRad*30.0,0);
>> InverseKinematics(LLEG_J5, Lfoot);

>> DrawAllJoints(1);  % Show the robot
```

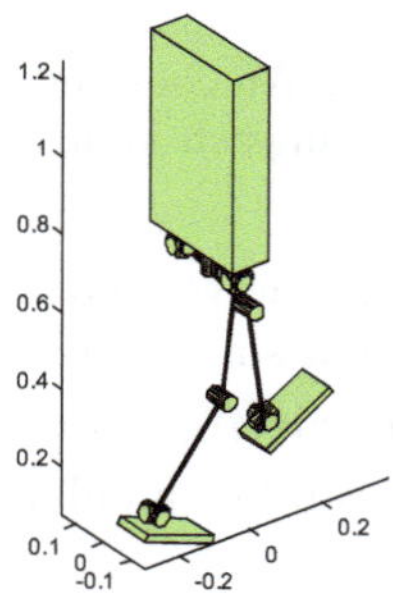

Fig. 2.29 Sample using InverseKinematics and Results of Calculation

2.5.5 Jacobian

In the previous section we introduced the Jacobian which gives you the relationship between small joint movements and spatial motion. Through the Jacobian we can also calculate the torque requirements of the joints in order to generate external forces through the hands and feet. As this is used extensively in robot control, papers on robotics that do not have a Jacobian somewhere in them are a rare thing indeed[19].

[19] We can find out how the Jacobian is used in robotics by reading Dr Yoshikawa's textbook [144].

Below we will go over the actual procedure of calculating the Jacobian. Figure 2.30 shows a chain with N links floating in space. We will assume that they are numbered in order from the base (link 1) to the end (link N). We will assume the end effector (robot hand or foot) is attached to the Nth link. Furthermore, we will assume that the forward kinematics have been calculated and the position and attitude of each link is already stored (jth link has $\boldsymbol{p}_j$, $\boldsymbol{R}_j$).

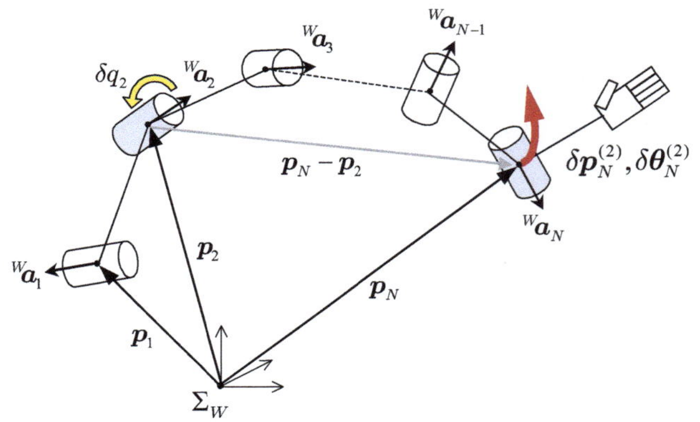

Fig. 2.30 Calculating the Jacobian: We consider the motion of the end effector that results from minute movements in each joint.

In this chain of links, let us assume we kept all the joints fixed except for the 2nd joint, which we turned by a small angle, δq_2. The amount of the end effector (link N)'s position changed($\delta\boldsymbol{p}_N^{(2)}$), and the amount the attitude of the end effector changed ($\delta\boldsymbol{\theta}_N^{(2)}$) can be calculated by

```
function J = CalcJacobian(idx)
global uLINK

jsize = length(idx);
target = uLINK(idx(end)).p;    % absolute target position
J = zeros(6,jsize);
for n=1:jsize
    j = idx(n);
    mom = uLINK(j).mother;
    a = uLINK(mom).R * uLINK(j).a;  % joint axis in world frame
    J(:,n) = [cross(a, target - uLINK(j).p) ; a ];
end
```

Fig. 2.31 CalcJacobian.m: Calculation of Jacobian

$$\begin{cases} \delta \boldsymbol{p}_N^{(2)} = {}^W\boldsymbol{a}_2 \times (\boldsymbol{p}_N - \boldsymbol{p}_2)\delta q_2 \\ \delta \boldsymbol{\theta}_N^{(2)} = {}^W\boldsymbol{a}_2 \delta q_2 \end{cases}$$

where ${}^W\boldsymbol{a}_2$ is the unit vector for the second joint axis with respect to the world frame

$${}^W\boldsymbol{a}_2 = \boldsymbol{R}_1 \boldsymbol{a}_2.$$

If we apply the same procedure on all the links from the 1st to the Nth and calculate their sum, we may obtain the change that occurs when all the joints are rotated by a small amount

$$\begin{cases} \delta \boldsymbol{p}_N = \sum_{j=1}^N \delta \boldsymbol{p}_N^{(j)} \\ \delta \boldsymbol{\theta}_N = \sum_{j=1}^N \delta \boldsymbol{\theta}_N^{(j)} \end{cases}. \tag{2.72}$$

We can rewrite the above as a matrix to obtain

$$\begin{bmatrix} \delta \boldsymbol{p}_N \\ \delta \boldsymbol{\theta}_N \end{bmatrix} = \begin{bmatrix} {}^W\boldsymbol{a}_1 \times (\boldsymbol{p}_N - \boldsymbol{p}_1) & \dots & {}^W\boldsymbol{a}_{N-1} \times (\boldsymbol{p}_N - \boldsymbol{p}_{N-1}) & 0 \\ {}^W\boldsymbol{a}_1 & \dots & {}^W\boldsymbol{a}_{N-1} & {}^W\boldsymbol{a}_N \end{bmatrix} \begin{bmatrix} \delta q_1 \\ \delta q_2 \\ \vdots \\ \delta q_N \end{bmatrix}. \tag{2.73}$$

In other words, the Jacobian $\boldsymbol{J}$ can be described by

$$\boldsymbol{J} \equiv \begin{bmatrix} {}^W\boldsymbol{a}_1 \times (\boldsymbol{p}_N - \boldsymbol{p}_1) & {}^W\boldsymbol{a}_2 \times (\boldsymbol{p}_N - \boldsymbol{p}_2) & \dots & {}^W\boldsymbol{a}_{N-1} \times (\boldsymbol{p}_N - \boldsymbol{p}_{N-1}) & 0 \\ {}^W\boldsymbol{a}_1 & {}^W\boldsymbol{a}_2 & \dots & {}^W\boldsymbol{a}_{N-1} & {}^W\boldsymbol{a}_N \end{bmatrix}. \tag{2.74}$$

This procedure implemented as a Matlab program is shown in Fig. 2.31.

2.5.6 Jacobian and the Joint Velocity

The relationship between the joint velocity and the end effector velocity is obtained by dividing (2.70) by a small time period δt

$$\frac{1}{\delta t} \begin{bmatrix} \delta \boldsymbol{p} \\ \delta \boldsymbol{\theta} \end{bmatrix} = \boldsymbol{J} \, \frac{\delta \boldsymbol{q}}{\delta t}.$$

Therefore, the joint speed $\dot{\boldsymbol{q}}$ and the end effector speed $(\boldsymbol{v}, \boldsymbol{\omega})$ are associated by the following

$$\begin{bmatrix} \boldsymbol{v} \\ \boldsymbol{\omega} \end{bmatrix} = \boldsymbol{J} \, \dot{\boldsymbol{q}}. \tag{2.75}$$

Note that this is the case when the body link is fixed in the space. If the body link has its own speed as $(\boldsymbol{v}_B, \boldsymbol{\omega}_B)$, we must use (see Section 2.3)

$$\begin{bmatrix} \boldsymbol{v} \\ \boldsymbol{\omega} \end{bmatrix} = \boldsymbol{J}\,\dot{\boldsymbol{q}} + \begin{bmatrix} \boldsymbol{v}_B + \boldsymbol{\omega}_B \times (\boldsymbol{p} - \boldsymbol{p}_B) \\ \boldsymbol{\omega}_B \end{bmatrix}. \tag{2.76}$$

In the following discussion, however, we assume the body link is fixed in the space for the sake of simplicity. We can calculate the joint speed which realize the tip speed $(\boldsymbol{v}, \boldsymbol{\omega})$ by the transformation of (2.75)

$$\dot{\boldsymbol{q}} = \boldsymbol{J}^{-1} \begin{bmatrix} \boldsymbol{v} \\ \boldsymbol{\omega} \end{bmatrix}. \tag{2.77}$$

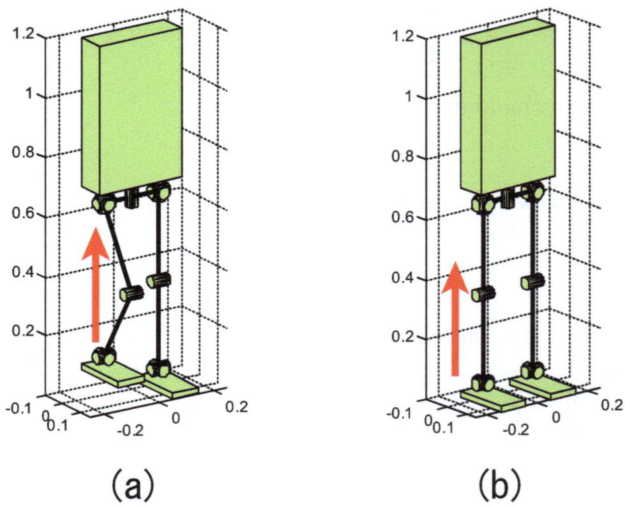

Fig. 2.32 Non-singular posture and singular posture

Let us calculate the joint speed for the two postures in Fig.2.32(a). First we set up all geometric information of the robot on Matlab command line, then use FindRoute() to find the path from the body to the right foot. Then the joint angles are set by using SetJointAngles() such that right hip pitch, knee pitch, and angle pitch angles to be $(-30, 60, -30)$ deg and 0 degs for other joints. We can calculate the Jacobian by the following inputs.

```
>> SetupBipedRobot;
>> idx = FindRoute(RLEG_J5);
>> SetJointAngles(idx,[0 0 -pi/6 pi/3 -pi/6 0])
>> J = CalcJacobian(idx);
```

Suppose we want to let the foot lift vertically with 0.1m/s, then the joint speeds can be obtained by (2.77).

```
>> dq = J \ [0 0 0.1 0 0 0]'
dq =
         0
         0
   -0.3333
    0.6667
   -0.3333
         0
```

From this result, we see the joint speeds for the hip pitch, knee pitch and ankle pitch are $(-0.3333, 0.6667, -0.3333)$ rad/s.

Let us try the same calculation for Fig.2.32(b) whose knee is fully stretched. For this purpose, we give zeros for all joint angles of right leg, calculate Jacobian, and obtain the joint speeds.

```
>> SetJointAngles(idx,[0 0 0 0 0 0])
>> J = CalcJacobian(idx);
>> dq = J \ [0 0 0.1 0 0 0]'
Warning: Matrix is singular to working precision.
dq =
   NaN
   NaN
   NaN
  -Inf
   Inf
     0
```

In this case, we have gotten an warning message "Matrix is singular to working precision." and obtained a joint speed vector consisting NaN (Not a number) and Inf (Infinity). Thus the calculation collapses at the singular pose of Fig.2.32(b), where the knee is fully stretched[20]. This happens because a robot can never move its foot vertically by applying any joint speeds at this configuration. A robot pose of such condition is called a **singular posture**. Let us observe what happens to an inverse calculation of the Jacobian at this singular posture.

```
>> J^(-1)
Warning: Matrix is singular to working precision.

ans =
   Inf   Inf   Inf   Inf   Inf   Inf
   Inf   Inf   Inf   Inf   Inf   Inf
   Inf   Inf   Inf   Inf   Inf   Inf
   Inf   Inf   Inf   Inf   Inf   Inf
   Inf   Inf   Inf   Inf   Inf   Inf
   Inf   Inf   Inf   Inf   Inf   Inf
```

[20] The robot joints can go out of control by illegal commands like NaN and Inf. Thus we must have warning and errors, but also take care to avoid implementing control commands with such illegal commands.

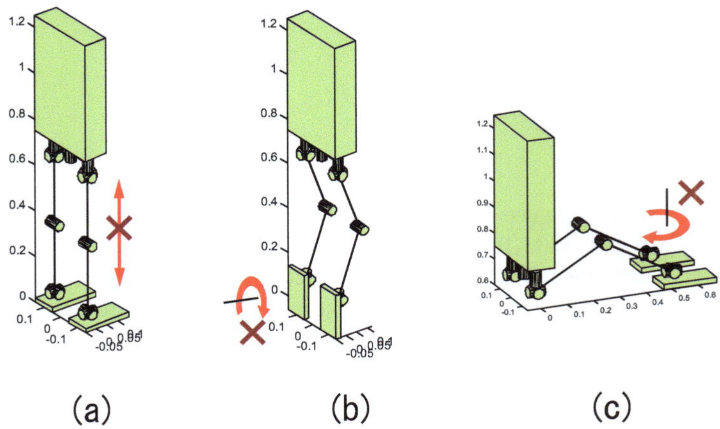

Fig. 2.33 Singular posture examples. In these postures there exist directions in which the end point cannot move (shown with arrows). In these cases the Jacobian inverse cannot be solved.

2.5.7 Singular Postures

Examples of singular postures are shown in Fig. 2.33. They are (a) the configuration with fully stretched knees as already explained in the previous subsection, (b) the pose where the hip yaw axis and the ankle roll axis are aligned, and (c) the pose where the hip roll axis and the ankle roll axis are aligned.

When a robot is in a singular posture, the Jacobian becomes a singular matrix and we cannot calculate its inverse. Let us check this for the case of Fig. 2.33(b).

```
>> SetupBipedRobot;
>> idx = FindRoute(RLEG_J5);
>> SetJointAngles(idx,[0 0 -pi/6 pi/3 pi/3 0])
>> J = CalcJacobian(idx);
>> J^(-1)
Warning: Matrix is close to singular or badly scaled.
         Results may be inaccurate. RCOND =  3.760455e-17B
ans =
   1.0e+15 *
          0   -8.7498          0    4.5465          0    0.0000
          0    0.0000          0    0.0000          0    0.0000
    -0.0000         0    -0.0000         0     0.0000         0
     0.0000         0     0.0000         0     0.0000         0
     0.0000         0    -0.0000         0     0.0000         0
          0   -8.7498          0    4.5465          0   -0.0000
>> det(J)
ans =
   8.9079e-18
```

```
>> rank(J)
ans =
     5
```

At the inverse calculation, we have got a warning message and a meaningless result with unrealistically huge ($\approx 10^{15}$) components due to numerical errors. From the same reason, the determinant is a very small value ($\approx 10^{-17}$) while it should be zero for a singular matrix. Finally, the rank of the Jacobian is five, which indicates the matrix singularity.

2.5.8 Inverse Kinematics with Singularity Robustness

The Newton-Raphson method of Section 2.5.4 does not work correctly around singular postures because of numerical instability. Let us observe it.

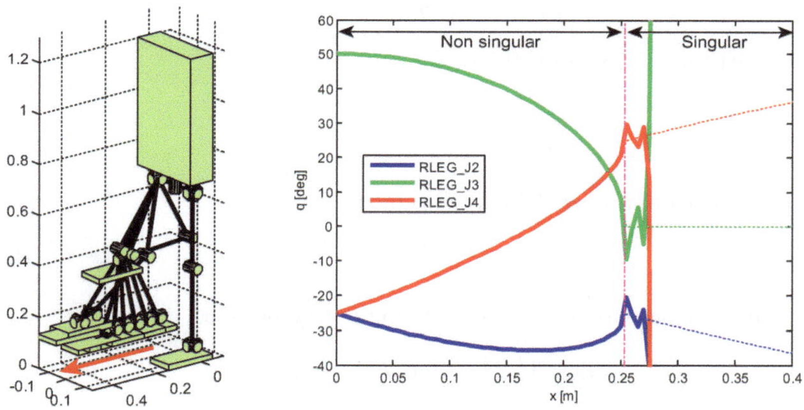

Fig. 2.34 Inverse kinematics by Newton-Raphson method around a singular configuration

In Fig. 2.34, joint angles were calculated for letting the right foot move forward from a non-singular pose by using Newton-Raphson method. The right graph shows the hip pitch, the knee pitch, and the ankle pitch angles at the given target foot position. Analytical solutions are shown by dotted lines for comparison. The foot position reached the singularity at the vertical chain line, then the numerical results started to vibrate and went off the chart (for example, the knee angle reaches over 8000 deg).

Now, let consider again the relationship between joint speed and end-effector velocity given by

$$\dot{\boldsymbol{x}} = \boldsymbol{J}\dot{\boldsymbol{q}}.$$

Here $\dot{\boldsymbol{x}}$ is a six dimension vector for the target velocity of the end-effector. Let us introduce an error vector $\boldsymbol{e}$ since this equality will no longer be satisfied at singularity.

$$\boldsymbol{e} := \dot{\boldsymbol{x}} - \boldsymbol{J}\dot{\boldsymbol{q}}. \tag{2.78}$$

We can always make the error $\boldsymbol{e}$ as small as possible, while it cannot reach zero in general. For this purpose, we define a cost function

$$E(\dot{\boldsymbol{q}}) = \frac{1}{2}\boldsymbol{e}^T\boldsymbol{e}. \tag{2.79}$$

If $\dot{\boldsymbol{q}}$ minimizes $E(\dot{\boldsymbol{q}})$, we will have,

$$\frac{\partial E(\dot{\boldsymbol{q}})}{\partial \dot{\boldsymbol{q}}} = 0. \tag{2.80}$$

By substituting (2.78) and (2.79), we get

$$\frac{\partial E(\dot{\boldsymbol{q}})}{\partial \dot{\boldsymbol{q}}} = -\boldsymbol{J}^T\dot{\boldsymbol{x}} + \boldsymbol{J}^T\boldsymbol{J}\dot{\boldsymbol{q}} = 0. \tag{2.81}$$

Therefore, $\dot{\boldsymbol{q}}$ to minimize $E(\dot{\boldsymbol{q}})$ is obtained by

$$\dot{\boldsymbol{q}} = (\boldsymbol{J}^T\boldsymbol{J})^{-1}\boldsymbol{J}^T\dot{\boldsymbol{x}}. \tag{2.82}$$

Unfortunately, we cannot use this at singularity because,

$$\det(\boldsymbol{J}^T\boldsymbol{J}) = \det(\boldsymbol{J}^T)\det(\boldsymbol{J}) = 0.$$

This means the inverse at the right side of (2.82) is not solvable.

Now, we slightly modify the cost function.

$$E(\dot{\boldsymbol{q}}) = \frac{1}{2}\boldsymbol{e}^T\boldsymbol{e} + \frac{\lambda}{2}\dot{\boldsymbol{q}}^T\dot{\boldsymbol{q}} \tag{2.83}$$

This evaluates the joint speed magnitude as well as the end-effector speed error. The joint speed is accounted by increasing the positive scalar λ. Again, by substituting (2.78) and (2.79), we get

$$\frac{\partial E(\dot{\boldsymbol{q}})}{\partial \dot{\boldsymbol{q}}} = -\boldsymbol{J}^T\dot{\boldsymbol{x}} + (\boldsymbol{J}^T\boldsymbol{J} + \lambda\boldsymbol{E})\dot{\boldsymbol{q}} = 0, \tag{2.84}$$

where $\boldsymbol{E}$ is an identity matrix of the same size of $\boldsymbol{J}^T\boldsymbol{J}$. The cost $E(\dot{\boldsymbol{q}})$ is minimized by

$$\dot{\boldsymbol{q}} = (\boldsymbol{J}^T\boldsymbol{J} + \lambda\boldsymbol{E})^{-1}\boldsymbol{J}^T\dot{\boldsymbol{x}}. \tag{2.85}$$

Even at singularity where $\det(\boldsymbol{J}) = 0$, we can always solve the above equation by using proper λ. Let us define a new matrix.

$$\begin{aligned}\dot{\boldsymbol{q}} &= \boldsymbol{J}^{\#\lambda}\dot{\boldsymbol{x}} \\ \boldsymbol{J}^{\#\lambda} &:= (\boldsymbol{J}^T\boldsymbol{J} + \lambda\boldsymbol{E})^{-1}\boldsymbol{J}^T. \end{aligned} \tag{2.86}$$

The matrix $\boldsymbol{J}^{\#\lambda}$ is called an SR inverse[21].

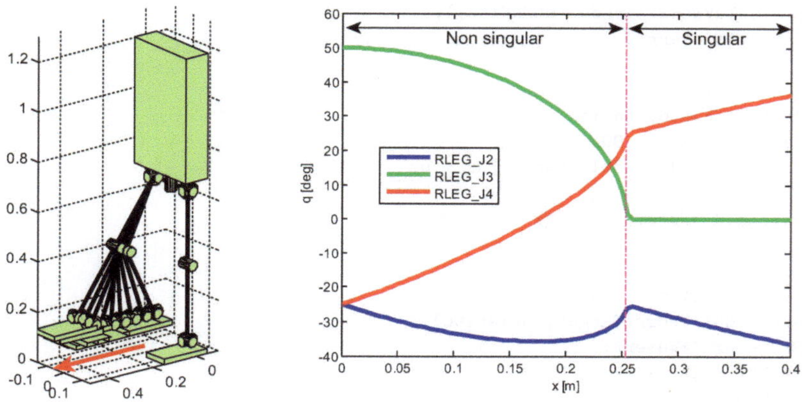

Fig. 2.35 Stable inverse kinematics calculation by SR inverse(Levenberg-Marquardt method)

By using SR inverse matrices, we can guarantee stable calculation of inverse kinematics at singularity. Figure 2.35 presents the robust and stable inverse kinematics calculation covering singularity and Fig. 2.36 shows its algorithm. To realize fast and robust calculation, the parameter λ should be modified based on the level of convergence. In this algorithm, we adopted a method proposed by Sugihara [124].

Similar optimization problems frequently appear in computer vision and machine learning. An iteration algorithm using SR inverse matrices are generally recognized as a version of Levenberg-Marquardt method [135].

2.5.9 Appendix: Supplementary Functions

The functions below are required in Section 2.5.4 and the succeeding subsections to calculate the numerical solution to the Inverse Kinematics and the Jacobian.

[21] SR stands for Singularity-Robust [138]. It is also called Damped Least-Square (DLS) Inverse [135].

```
function err_norm = InverseKinematics_LM(to, Target)
global uLINK

idx = FindRoute(to);
wn_pos = 1/0.3;  wn_ang = 1/(2*pi);
We = diag([wn_pos wn_pos wn_pos wn_ang wn_ang wn_ang]);
Wn = eye(length(idx));

ForwardKinematics(1);
err = CalcVWerr(Target, uLINK(to));
Ek = err'*We*err;

for n = 1:10
  J  = CalcJacobian(idx);
  lambda = Ek + 0.002;
  Jh = J'*We*J + Wn*lambda;  %Hk + wn

  gerr = J'*We*err;     %gk
  dq   = Jh \ gerr;     %new

  MoveJoints(idx, dq);
  err = CalcVWerr(Target, uLINK(to));
  Ek2 = err'*We*err;
  if Ek2 < 1E-12
     break;
  elseif Ek2 < Ek
      Ek = Ek2;
  else
      MoveJoints(idx, -dq);  % revert
      break,
  end
end
```

Fig. 2.36 InvserseKinematics_LM.m Inverse kinematics algorithm to handle singularities

```
function idx = FindRoute(to)
global uLINK

i = uLINK(to).mother;
if i == 1
    idx = [to];
else
    idx = [FindRoute(i) to];
end
```

Fig. 2.37 FindRoute.m Find a route from the body to the target link

```
function SetJointAngles(idx, q)
global uLINK

for n=1:length(idx)
    j = idx(n);
    uLINK(j).q = q(n);
end
ForwardKinematics(1);
```

Fig. 2.38 SetJointAngles.m Set joint angles along the specified indexes

```
function err = CalcVWerr(Cref, Cnow)

perr = Cref.p - Cnow.p;
Rerr = Cnow.R^-1 * Cref.R;
werr = Cnow.R * rot2omega(Rerr);
err = [perr; werr];
```

Fig. 2.39 CalcVWerr.m Function to calculate the amount of error in the position and attitude

```
function w = rot2omega(R)

el = [R(3,2)-R(2,3); R(1,3)-R(3,1); R(2,1)-R(1,2)];
norm_el = norm(el);
if norm_el > eps
    w = atan2(norm_el, trace(R)-1)/norm_el * el;
elseif R(1,1)>0 && R(2,2)>0 && R(3,3)>0
    w = [0 0 0]';
else
    w = pi/2*[R(1,1)+1; R(2,2)+1; R(3,3)+1];
end
```

Fig. 2.40 rot2omega.m Transform rotation matrix into the corresponding angular velocity vector (2.39)

Chapter 3
ZMP and Dynamics

The main topic of this section is the physics of the robot while that of the foregoing chapter being the geometry.

We first show a method for measuring the ZMP which is an important physical quantity for humanoid robots. Then we show a method for calculating the ZMP for a given motion of a humanoid robot. Lastly, we explain a certain mistake on the ZMP and cases which cannot be handled by using the ZMP.

3.1 ZMP and Ground Reaction Forces

The base of an industrial robot is fixed to the ground while the sole of a humanoid robot is not fixed and just contacts with the ground. Because of this, although the industrial robots can move freely within the joint movable range, the humanoid robot has to move with keeping the difficult condition of maintaining contact between the sole and the ground. Here, given a motion of a humanoid robot, we need to judge whether or not the contact can be maintained between the sole and the ground. Also, we need to plan a motion of a humanoid robot maintaining contact between the sole and the ground. We usually use the ZMP for these kinds of purposes.

3.1.1 ZMP Overview

1 Definition of ZMP

In 1972, Vukobratović and Stepanenko defined the **Zero-Moment Point (ZMP)** at the beginning of the paper on control of humanoid robots[1]. Everything of the argument regarding the ZMP starts from here.

[1] We can find the same definition in the book [81]. Later, some delicate aspects of the ZMP definition were discussed by Vukobratović and Borovac [88].

S. Kajita et al., *Introduction to Humanoid Robotics*, Springer Tracts in Advanced Robotics 101,
DOI: 10.1007/978-3-642-54536-8_3,

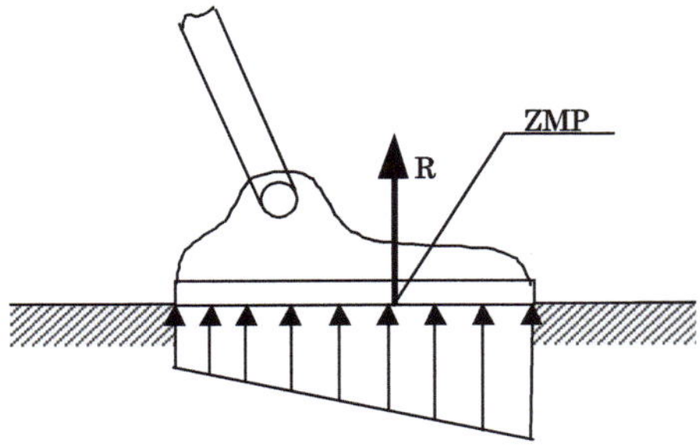

Fig. 3.1 Definition of Zero-Moment Point (ZMP) [90]

In Fig. 3.1 an example of force distribution across the foot is given. As the load has the same sign all over the surface, it can be reduced to the resultant force R, the point of attack of which will be in the boundaries of the foot. Let the point on the surface of the foot, where the resultant R passed, be denoted as the zero-moment point, or ZMP in short.

2 ZMP and Support Polygon

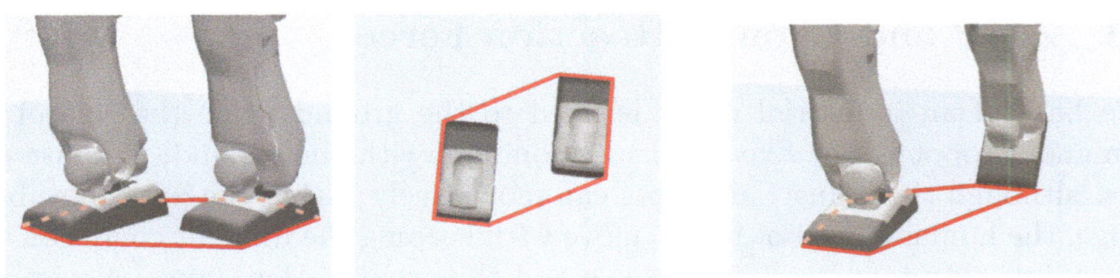

(a) Full contact of both feet (b) Partial contact

Fig. 3.2 Support Polygon

We explain the support polygon which is another important concept related to the ZMP. As shown in Fig. 3.2, let us consider the region formed by enclosing all the contact points between the robot and the ground by using an elastic cord braid. We call this region as the **support polygon**. Mathematically the support polygon is defined as a convex hull, which is the smallest convex set including all contact points. Definitions of the convex set and the convex hull are explained in the appendix of this chapter.

Rather than detailed discussions, we first show a simple and important relationship between the ZMP and the support polygon, i.e.,

The ZMP always exists inside of the support polygon.

Here, Vukovratović originally stated "the point which exists inside the boundary of the foot." In order to illustrate this more concretely, consider the images in Fig.3.3 illustrating the relationship among the center of mass (CoM), ZMP and the support polygon while a human stands on the ground. We call **the ground projection of CoM** the point where the gravity line from the CoM intersects the ground. As shown in Fig. 3.3(a), when a human stands on the ground, the ZMP coincides with the ground projection of CoM. In such a case, a human can keep balance if the ground projection of CoM is included strictly inside of the support polygon. On the other hand, when a human moves dynamically as shown in Fig. 3.3(b), the ground projection of CoM may exist outside the support polygon. However, the ZMP never exists outside the support polygon. In the following, we will explain the reason why the ZMP is always included in the support polygon.

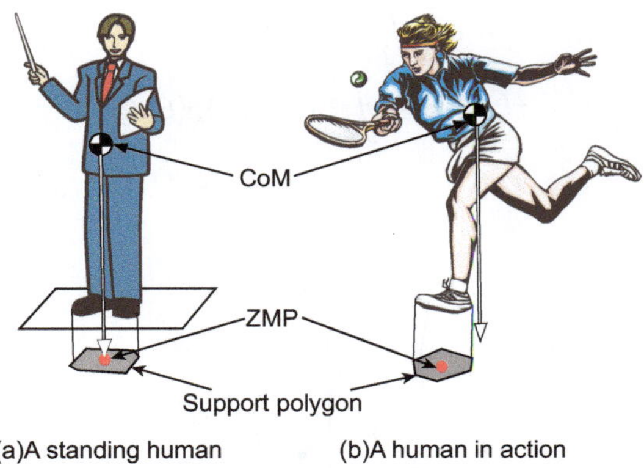

Fig. 3.3 CoG, ZMP, and Support Polygon

3.1.2 2D Analysis

1 ZMP in 2D

In Fig. 3.1, although only the vertical component of the ground reaction force is shown, the horizontal component of it also exists due to the friction between the ground and the soles of the feet.

In Fig. 3.4(a) and (b), we separately show the vertical component$\rho(\xi)$ and the horizontal component $\sigma(\xi)$ of the ground reaction force per unit length of the sole. These forces simultaneously act on a humanoid robot.

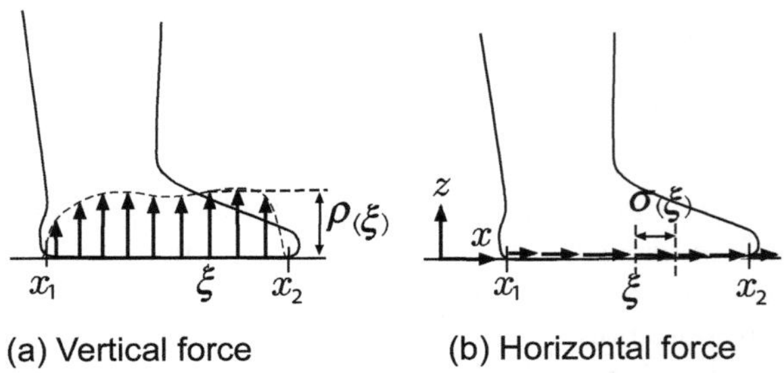

Fig. 3.4 Ground Reaction Force of 2D Model

Let us replace the forces distributed over the sole by a equivalent force and moment at a certain point in the sole. The force (f_x and f_z) and the moment ($\tau(p_x)$) at the point p_x in the sole can be expressed by

$$f_x = \int_{x_1}^{x_2} \sigma(\xi) d\xi \tag{3.1}$$

$$f_z = \int_{x_1}^{x_2} \rho(\xi) d\xi \tag{3.2}$$

$$\tau(p_x) = -\int_{x_1}^{x_2} (\xi - p_x)\rho(\xi) d\xi. \tag{3.3}$$

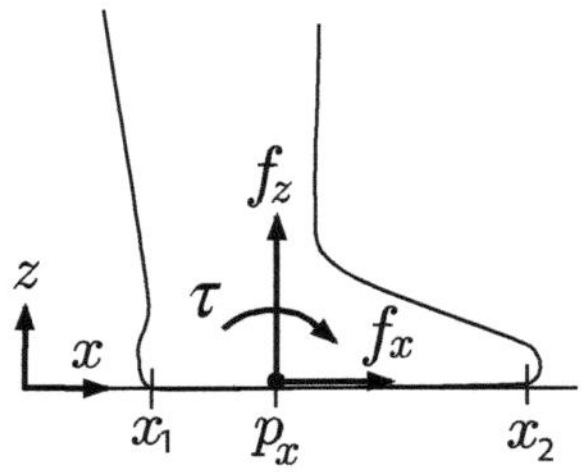

Fig. 3.5 Ground Reaction Forces and their Equivalent Force and Moment

Focusing on (3.3) with respect to the moment, let us consider the point p_x where moment becomes zero. Considering $\tau(p_x) = 0$ for (3.3), p_x can be obtained as follows:

$$p_x = \frac{\int_{x_1}^{x_2} \xi \rho(\xi) d\xi}{\int_{x_1}^{x_2} \rho(\xi) d\xi}. \tag{3.4}$$

Here, $\rho(\xi)$ is equivalent to the pressure since it is the vertical component of forces per unit length. Thus, p_x defined in (3.4) is the **center of pressure** and is the ZMP defined in the previous section. For 2D cases, since the ZMP is a point where **the moment of the ground reaction force becomes zero**, it has become the origin of the name.

2 Region of ZMP in 2D

Generally speaking, since the vertical component of the ground reaction force does not become negative unless magnets or suction cups are attached at the sole,

$$\rho(\xi) \geq 0.$$

Substituting this relationship into (3.4), we obtain

$$x_1 \leq p_x \leq x_2. \tag{3.5}$$

Equation (3.5) means that the ZMP is included in the segment of contact between the sole and the ground and does not exist outside it.

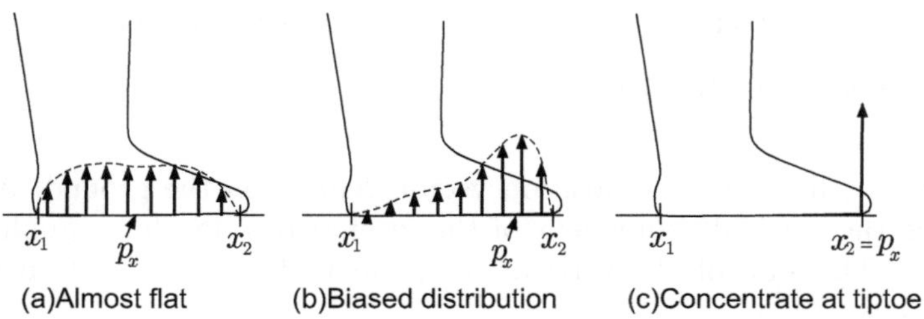

Fig. 3.6 ZMP and Pressure Distribution

Figure 3.6 shows the relationship between the pressure distribution and the position of the ZMP. As shown in (a), when the reaction force is distributed over the sole almost equally, the ZMP exists at the center of the sole. On the other hand, as shown in (b), when the distribution is inclined to the front part of the sole, the ZMP exists at the front part of the sole. Furthermore, as shown in (c), when a point at the toe supports all the reaction forces, the ZMP also exists at the toe. In this case, since the surface contact between the sole and ground is not guaranteed any longer, the foot begins to rotate about the toe by only a slight external disturbance applied to the robot. To reduce the danger of falling down when a humanoid robot moves, it is desirable to have the ZMP located inside of the support polygon while maintaining a certain margin from the end.

3.1.3 3D Analysis

We now extend the concept of the ZMP to 3D cases.

1 Ground Reaction Force in 3D

Let us consider the ground reaction force applied to the robot moving in 3D space from the flat ground. The horizontal component and the vertical component of the ground reaction forces are shown in Figs. 3.7(a) and (b), respectively. In actual situations, the sum of these two components is applied to the robot at the same time.

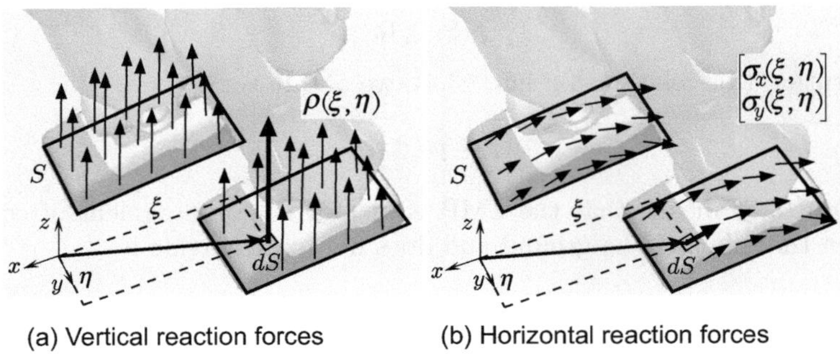

Fig. 3.7 Ground Reaction Force in 3D

Let $\boldsymbol{r} = [\xi \ \eta \ 0]^T$ be the position vector defined on the ground. Also, let $\rho(\xi, \eta)$ be the vertical component of the ground reaction force applied at a unit area. The sum of the vertical component of ground reaction force is expressed as

$$f_z = \int_S \rho(\xi, \eta) dS, \tag{3.6}$$

where $\int_S$ denotes the area integration at the contact between the sole and the ground. The moment $\boldsymbol{\tau}_n(\boldsymbol{p})$ of the ground reaction force about a point $\boldsymbol{p} = [p_x \ p_y \ 0]^T$ can be calculated as

$$\boldsymbol{\tau}_n(\boldsymbol{p}) \equiv [\tau_{nx} \ \tau_{ny} \ \tau_{nz}]^T \tag{3.7}$$

$$\tau_{nx} = \int_S (\eta - p_y) \rho(\xi, \eta) dS \tag{3.8}$$

$$\tau_{ny} = -\int_S (\xi - p_x) \rho(\xi, \eta) dS \tag{3.9}$$

$$\tau_{nz} = 0.$$

As well as the 2D cases, assuming

$$\tau_{nx} = 0 \tag{3.10}$$

$$\tau_{ny} = 0 \tag{3.11}$$

for (3.8) and (3.9), the point where the moment of the vertical component of the ground reaction force becomes zero can be expressed as

$$p_x = \frac{\int_S \xi\rho(\xi,\eta)dS}{\int_S \rho(\xi,\eta)dS} \tag{3.12}$$

$$p_y = \frac{\int_S \eta\rho(\xi,\eta)dS}{\int_S \rho(\xi,\eta)dS}. \tag{3.13}$$

Since $\rho(\xi,\eta)$ is equivalent to the pressure over the surface of the sole, the point $\boldsymbol{p}$ is the **center of perssure** or in other word, ZMP.

On the other hand, let us consider the effect of the horizontal component of the ground reaction force. Let $\sigma_x(\xi,\eta)$ and $\sigma_y(\xi,\eta)$ be the x and y components, respectively, of the horizontal ground reaction forces. The sum of them can be expressed as

$$f_x = \int_S \sigma_x(\xi,\eta)dS \tag{3.14}$$

$$f_y = \int_S \sigma_y(\xi,\eta)dS. \tag{3.15}$$

The moment $\boldsymbol{\tau}_t(\boldsymbol{p})$ of the horizontal ground reaction force about a point $\boldsymbol{p}$ on the ground surface is expressed as

$$\begin{aligned}
\boldsymbol{\tau}_t(\boldsymbol{p}) &\equiv [\tau_{tx}\ \tau_{ty}\ \tau_{tz}]^T \\
\tau_{tx} &= 0 \\
\tau_{ty} &= 0 \\
\tau_{tz} &= \int_S \{(\xi - p_x)\sigma_y(\xi,\eta) - (\eta - p_y)\sigma_x(\xi,\eta)\}dS.
\end{aligned} \tag{3.16}$$

These equations mean that the horizontal ground reaction forces generate the vertical component of the moment.

From the above discussions, we can see that, as shown in Fig. 3.8, the ground reaction forces distributed over the surface of the sole can be replaced by the force

$$\boldsymbol{f} = [f_x\ f_y\ f_z]^T,$$

and the moment

$$\begin{aligned}
\boldsymbol{\tau}_p &= \boldsymbol{\tau}_n(\boldsymbol{p}) + \boldsymbol{\tau}_t(\boldsymbol{p}) \\
&= [0\ 0\ \tau_{tz}]^T,
\end{aligned}$$

about the ZMP $\boldsymbol{p}$. When a robot moves, $\tau_{tz} = 0$ is not generally satisfied. Thus, the ZMP is not a point where all components of the moment becomes zero for 3D cases. The ZMP is defined as **the point where the horizontal component of the moment of the ground reaction forces becomes zero** for 3D cases.

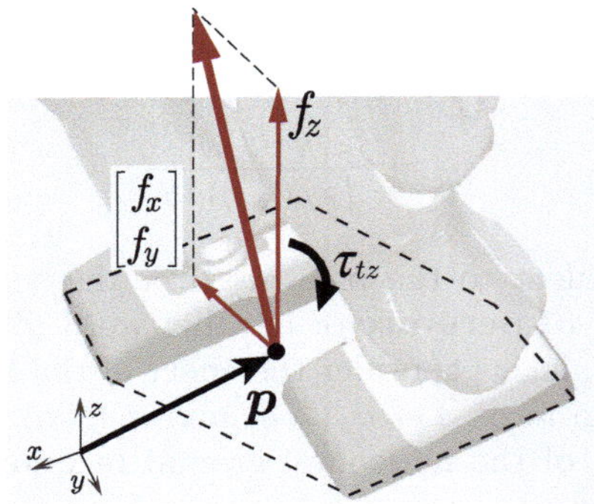

Fig. 3.8 Ground Reaction Forces and Equivalent Force and Moment for 3D Models

2 Region of ZMP in 3D

Let us define the region of the ZMP for 3D cases. For simplicity, we consider the ground reaction forces $\boldsymbol{f}_i = [f_{ix}\ f_{iy}\ f_{iz}]^T$ acting at the discretized points $\boldsymbol{p}_i \in S$ $(i = 1, \cdots, N)$ as shown in Fig. 3.9. This approximation becomes more exact as the number of discretized points increases.

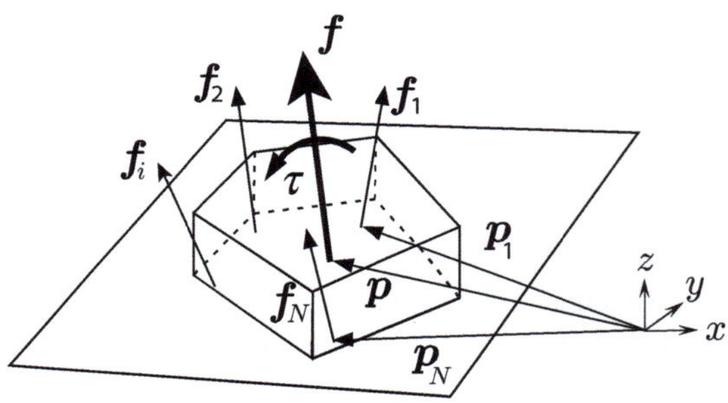

Fig. 3.9 Force/Moment at the ZMP Expressed by Forces at Discretized Points

Next, distributed N force vectors are replaced by a force and a moment vectors acting at the point $\boldsymbol{p}$ as

$$\boldsymbol{f} = \sum_{i=1}^{N} \boldsymbol{f}_i \tag{3.17}$$

$$\boldsymbol{\tau}(\boldsymbol{p}) = \sum_{i=1}^{N} (\boldsymbol{p}_i - \boldsymbol{p}) \times \boldsymbol{f}_i. \tag{3.18}$$

The position of the ZMP can be obtained by setting the first and the second elements of (3.18) be zero. This yields

$$p = \frac{\sum_{i=1}^{N} p_i f_{iz}}{\sum_{i=1}^{N} f_{iz}}. \tag{3.19}$$

For ordinary humanoid robots without magnets or suction cups at the sole, the vertical component of the ground reaction forces becomes zero for all discretized points, i.e.,

$$f_{iz} \geq 0 \qquad (i = 1, \cdots, N). \tag{3.20}$$

Here, introducing the new variables $\alpha_i = f_{iz} / \sum_{j=1}^{N} f_{jz}$, we obtain

$$\begin{cases} \alpha_i \geq 0 & (i = 1, \cdots, N) \\ \sum_{i=1}^{N} \alpha_i = 1. \end{cases} \tag{3.21}$$

Rewriting (3.19) by using α_i, the region of the ZMP can be expressed as

$$p \in \left\{ \sum_{i=1}^{N} \alpha_i p_i \,\middle|\, p_i \in S \ (i = 1, \cdots, N) \right\}. \tag{3.22}$$

By comparing (3.21) and (3.22) with the definition of convex hull (3.90) in section 3.6, we can see that the ZMP is included in the convex hull of the set S, in other words, the support polygon.

3.2 Measurement of ZMP

This section explains methods for measuring the position of the ZMP by using several sensors attached at the feet of a humanoid robot. For a biped walking robots to measure the position of the ZMP, we should consider two cases, i.e., (1) **the ZMP of each foot** considering the reaction force between either one of the feet and the ground, and (2) the ZMP considering the reaction force between both feet and the ground. During the double support phase, these two ZMPs becomes different.

3.2.1 General Discussion

Let us consider the model shown in Fig. 3.10. In this model, there are two rigid bodies contacting each other where one of them also contacts the ground. The forces and moments applied by one rigid body to the other are measured at multiple points. This model imitates the foot of a humanoid robot. When the robot moves and the foot is forced on the ground, the output of the force/torque sensor at the foot is generated. By using this sensor information, the position of the ZMP is measured.

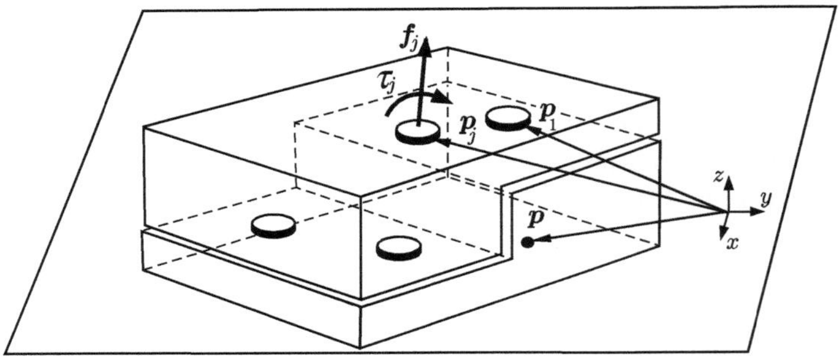

Fig. 3.10 Definition of Variables with respect to the Position and the Output of Force/Torque Sensors

Let us assume that, at the points $\boldsymbol{p}_j$ $(j = 1, \cdots, N)$ with respect to the reference coordinate system, the forces $\boldsymbol{f}_j$ and moments $\boldsymbol{\tau}_j$ are measured. Here, the moment about the point $\boldsymbol{p} = [p_x \; p_y \; p_z]^T$ is

$$\boldsymbol{\tau}(\boldsymbol{p}) = \sum_{j=1}^{N} (\boldsymbol{p}_j - \boldsymbol{p}) \times \boldsymbol{f}_j + \boldsymbol{\tau}_j. \tag{3.23}$$

The position of the ZMP can be obtained by setting the x and y components of (3.23) be zero and by solving for p_x and p_y as

$$p_x = \frac{\sum_{j=1}^{N} \{-\tau_{jy} - (p_{jz} - p_z) f_{jx} + p_{jx} f_{jz}\}}{\sum_{j=1}^{N} f_{jz}} \tag{3.24}$$

$$p_y = \frac{\sum_{j=1}^{N} \{\tau_{jx} - (p_{jz} - p_z) f_{jy} + p_{jy} f_{jz}\}}{\sum_{j=1}^{N} f_{jz}} \tag{3.25}$$

where

$$\boldsymbol{f}_j = [f_{jx} \; f_{jy} \; f_{jz}]^T$$
$$\boldsymbol{\tau}_j = [\tau_{jx} \; \tau_{jy} \; \tau_{jz}]^T$$
$$\boldsymbol{p}_j = [p_{jx} \; p_{jy} \; p_{jz}]^T .$$

Equations (3.24) and (3.25) are the basis for measuring the position of the ZMP[2].

[2] When a foot does not contact the ground, the ZMP position cannot be determined since the denominators of (3.24) and (3.25) become zero. Therefore, when measuring the ZMP, we have to introduce a threshold and set $p_x = p_y = 0$ when the denominator is less than the threshold.

3.2.2 ZMP of Each Foot

First, focusing on the contact between one foot and the ground, we measure the ZMP.

1 Measurement Using 6 Axis Force/Torque Sensor

Figure 3.11 shows the foot of the humanoid robot HRP-2 [65]. The ground reaction force applied to the sole is transmitted to the sensor mount through rubber bushes and dampers. A 6 axis force/torque sensor is attached at the sensor mount, and the force is transmitted to the ankle of the robot through this sensor. The rubber bushes and the dampers are positioned to prevent large impulse forces from being transmitted to the robot. Since the displacement of them is small, we do not consider the displacement when calculating the ZMP.

A 6 axis force/torque sensor is coordinated to simultaneously measure the force $\boldsymbol{f} = [f_x, f_y, f_z]$ and the moment $\boldsymbol{\tau} = [\tau_x\ \tau_y\ \tau_z]$ applied from outside the robot. This sensor is mainly used for measuring the force at the end effector of industrial robots. An example of 6 axis force/torque sensor is shown in Fig. 3.12. To measure the ZMP of a humanoid robot, the force/torque sensor must be light and must be strong enough to accept the large impulsive force applied to the sensor.

To obtain the ZMP from the measured data of 6 axis force/torque sensor, we set $N = 1$ in (3.24) and (3.25).

Let the position of the ZMP in the right and the left foot be $\boldsymbol{p}_R$ and $\boldsymbol{p}_L$, respectively, as shown in Fig. 3.13. Especially when the center of measurement

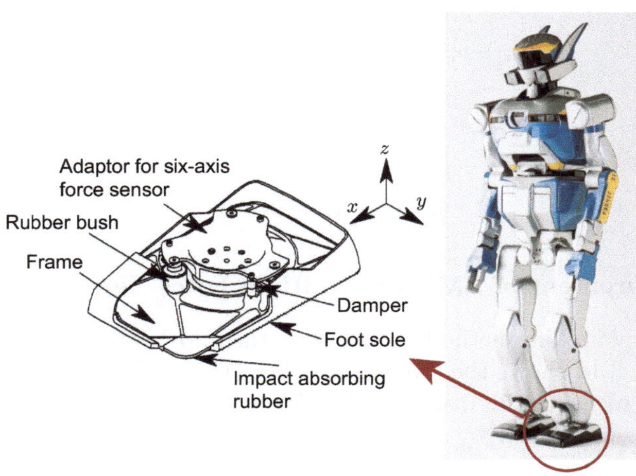

Fig. 3.11 Foot of HRP-2 [65].

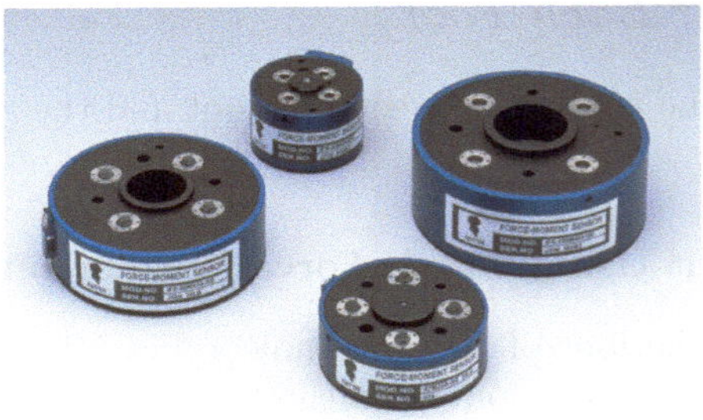

Fig. 3.12 An Example of 6 Axis Force/Torque Sensor (courtesy of Nitta Corp.)

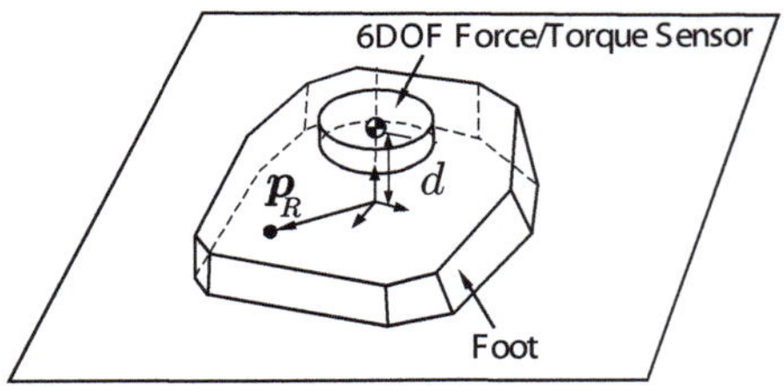

Fig. 3.13 Definition of Variables for Calculation of ZMP by 6 Axis Force/Torque Sensor

of the sensor lies on the z axis of the reference coordinate system, the position of the ZMP of each foot can be obtained very simply. For the right foot,

$$p_{Rx} = (-\tau_{1y} - f_{1x}d)/f_{1z} \tag{3.26}$$
$$p_{Ry} = (\tau_{1x} - f_{1y}d)/f_{1z} \tag{3.27}$$

where

$$\boldsymbol{p}_R = [p_{Rx}\ p_{Ry}\ p_{Rz}]^T$$
$$\boldsymbol{p}_1 = [0\ \ 0\ \ d]^T.$$

2 Measurement of ZMP by Multiple Force Sensors

Next, we explain the method to measure the ZMP by using multiple force sensors. Fig. 3.14 shows the humanoid robot H5 [70]. To make the foot light, the ZMP is measured by using twelve **force sensing registers: FSR** and sorbothane sandwiched by two aluminum planes (Fig. 3.14(b)). Since the electric resistance changes according to the applied force, the FSR can be used as a **one dimensional force sensor** to measure the vertical component of ground reaction force.

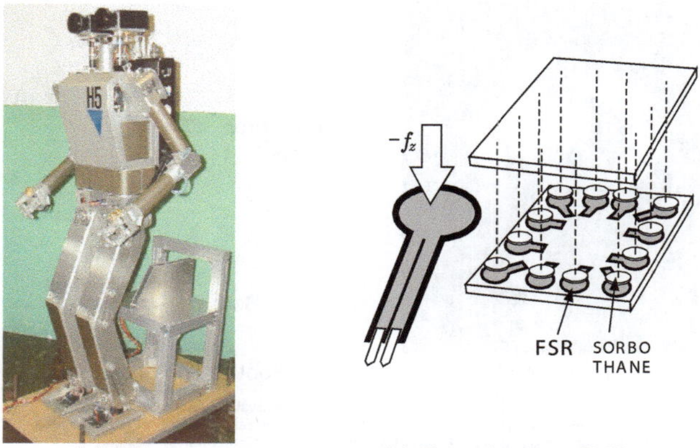

(a) H5 (b) Foot design of H5

Fig. 3.14 Humanoid Robot H5 and its Foot
(Courtesy of Dept. of Mechano-Informatics, The Univ. of Tokyo)

To measure the ZMP, the x and y components of the force are set to be 0 in (3.24) and (3.25). As shown in Fig.3.15, when there are N one-dimensional force sensors, the ZMP can be obtained by

$$p_x = \frac{\sum_{j=1}^{N} p_{jx} f_{jz}}{\sum_{j=1}^{N} f_{jz}} \tag{3.28}$$

$$p_y = \frac{\sum_{j=1}^{N} p_{jy} f_{jz}}{\sum_{j=1}^{N} f_{jz}}. \tag{3.29}$$

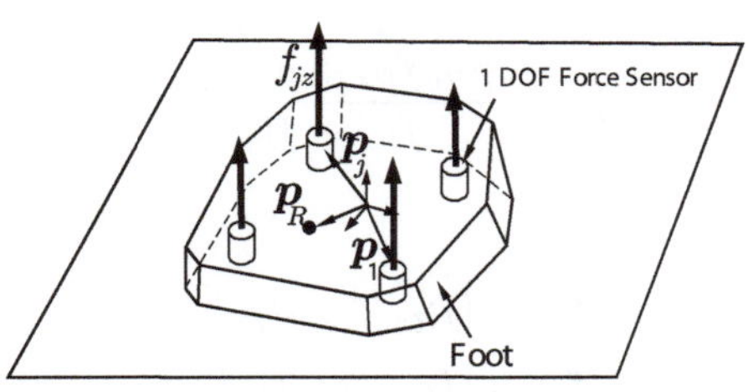

Fig. 3.15 Definition of Variables for Calculation of ZMP by Multiple 1 Axis Force Sensor

Figure 3.16 shows the humanoid robot Morph3 and its foot [129, 120]. Morph3 measures the ZMP by using four **3 axis force sensors** attached at each foot (Fig. 3.16(b)). The 3 axis force sensor measures the 3 dimensional forces applied to the sole split into four parts. By using this measuring system, we can obtain measurement on the point of contact. To calculate the ZMP

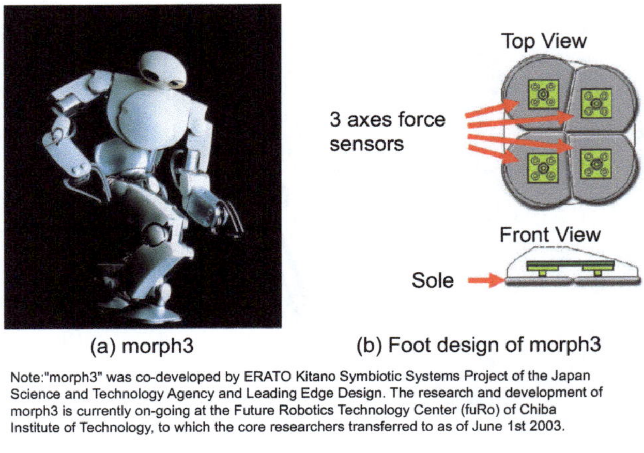

Fig. 3.16 Humanoid Robot Morph3 and its Foot

of each foot, the element of moment, τ_{jx} and τ_{jy}, in (3.24) and (3.25) are set to be zero.

3.2.3 ZMP for Both Feet Contact

Until the previous section, the position of the ZMP of each foot can be obtained as $\boldsymbol{p}_R$ and $\boldsymbol{p}_L$. The ground reaction forces $\boldsymbol{f}_R$ and $\boldsymbol{f}_L$ are also obtained from the sensor information. By using this information, we calculate the ZMP in the case where both feet are in contact with the ground. By using (3.24) and (3.25), the ZMP can be obtained as

$$p_x = \frac{p_{Rx} f_{Rz} + p_{Lx} f_{Lz}}{f_{Rz} + f_{Lz}} \tag{3.30}$$

$$p_y = \frac{p_{Ry} f_{Rz} + p_{Ly} f_{Lz}}{f_{Rz} + f_{Lz}} \tag{3.31}$$

where

$$\boldsymbol{f}_R = [f_{Rx}\ f_{Ry}\ f_{Rz}]^T$$
$$\boldsymbol{f}_L = [f_{Lx}\ f_{Ly}\ f_{Lz}]^T$$
$$\boldsymbol{p}_R = [p_{Rx}\ p_{Ry}\ p_{Rz}]^T$$
$$\boldsymbol{p}_L = [p_{Lx}\ p_{Ly}\ p_{Lz}]^T .$$

During the single support phase, since the vertical component of the ground reaction force becomes zero, the ZMP calculated using (3.30) and (3.31) coincides with the ZMP of the supporting foot. This yields

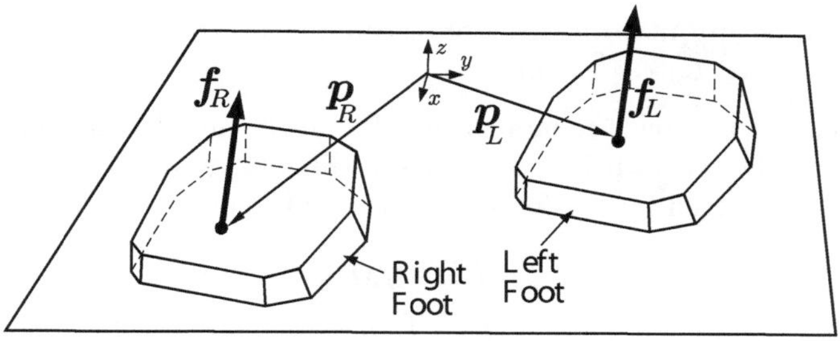

Fig. 3.17 Definition of Variables for ZMP for Both Feet Contact

$$[p_x\ p_y\ p_z]^T = \begin{cases} [p_{Rx}\ p_{Ry}\ p_{Rz}]^T & \text{for support of right foot} \\ [p_{Lx}\ p_{Ly}\ p_{Lz}]^T & \text{for support of left foot.} \end{cases} \quad (3.32)$$

We conclude this section stating that, when we consider the balance of a humanoid robot, we can use (3.30) and (3.31) taking the both feet into account regardless of the supporting foot.

3.3 Dynamics of Humanoid Robots

From the previous discussion, we can express the ground reaction force acting upon a humanoid robot by using the ZMP, the linear force, and the moment about a vertical line passing the ZMP. In this section, we discuss the relationship between the ground reaction force and the robot's motion. After showing basic equations, we explain the principle of it. Lastly, we show some calculation algorithms.

3.3.1 Humanoid Robot Motion and Ground Reaction Force

1 Basic Physical Parameters

Let us consider a humanoid robot with arbitrary structure. While it can be composed of metal, plastic, and ceramics etc., we assume that we can clearly identify between the robot and other things. We can define the following ten physical parameters classified into four groups:

Mass: Total Robot's mass. M [kg]
Center of Mass: Robot's center of mass. $\boldsymbol{c} \equiv [x\ y\ z]^T$ [m]
Momentum: Measure of an robot's translational motion[3].
$\mathcal{P} \equiv [\mathcal{P}_x\ \mathcal{P}_y\ \mathcal{P}_z]^T$ [Ns]

[3] We often call it the linear momentum to distinguish it from the angular momentum.

Angular Momentum: Measure of robot's rotational motion about the origin. $\mathcal{L} \equiv [\mathcal{L}_x \ \mathcal{L}_y \ \mathcal{L}_z]^T$ [Nms]

We will make clear what the momentum $\mathcal{P}$ and the angular momentum $\mathcal{L}$ are subsequently. The **dynamics** gives laws to these physical quantities. And, its principle is expressed by the following three equations:

$$\begin{aligned} \dot{\boldsymbol{c}} &= \mathcal{P}/M \\ \dot{\mathcal{P}} &= \boldsymbol{f}_{all} \\ \dot{\mathcal{L}} &= \boldsymbol{\tau}_{all}. \end{aligned}$$

We will explain the meaning of each equation.

2 Dynamics of Translational Motion

The first equation with respect to the translational motion gives the relationship between the velocity of the mass center and the momentum

$$\dot{\boldsymbol{c}} = \mathcal{P}/M. \tag{3.33}$$

Conversely, we can see from this equation that the momentum is (total mass)×(velocity of mass center). The second equation with respect to the translational motion shows how the momentum changes according to the external forces

$$\dot{\mathcal{P}} = \boldsymbol{f}_{all}. \tag{3.34}$$

where $\boldsymbol{f}_{all}$ denotes the sum of the forces applied to the robot from outside it. Since the unit of force is [N], we can see from this equation that the unit of momentum is [Ns][4]. Newton's second law of motion described in "Principia Mathematica Philosophiae Naturalis" was originally expressed in the style of (3.34). The familiar relationship between the force and acceleration can be obtained by deleting $\mathcal{P}$ from (3.34) and (3.33).

Let us consider the force applied from outside the robot. The gravitational force is equally applied to all the components of the robot, and its sum can be regarded as the force $M\boldsymbol{g}$ applied at the robot's center of mass $\boldsymbol{c}$. Here, $\boldsymbol{g}$ denotes the **gravitational acceleration** vector where it is $[0\ 0\ -9.8]^T$ [m/s^2] on the Earth, $[0\ 0\ -1.7]^T$ on the Moon, and and $[0\ 0\ -3.6]^T$ on Mars. Since the gravitational force is always applied regardless of the robot's motion, it is treated separately from other forces,

$$\boldsymbol{f}_{all} = M\boldsymbol{g} + \boldsymbol{f}.$$

[4] Let's confirm that it is coincides with the unit [kgm/s] obtained by mass×velocity.

While $\boldsymbol{f}$ denotes the force other than the gravity, we consider it as the ground reaction forces[5]. Therefore, the equation of translational motion can be obtained by

$$\dot{\mathcal{P}} = M\boldsymbol{g} + \boldsymbol{f}. \tag{3.35}$$

When a robot stands still, the change of momentum is 0 since the gravitational force balances with the ground reaction force. If the ground reaction force disappears, the robot's momentum increases rapidly downward due to gravity. This is **free fall.**

3 Dynamics of Rotational Motion

We have the following equation with respect to the rotational motion

$$\dot{\mathcal{L}} = \boldsymbol{\tau}_{all}. \tag{3.36}$$

This equation shows that the angular momentum changes according to the sum of the moments, $\boldsymbol{\tau}_{all}$ applied from the outside the robot. Since the unit of the moment is [Nm], we can see from this equation that the unit of angular momentum is [Nms].

Among the moments applied to the robot, the moment generated by the gravitational force is given by

$$\boldsymbol{\tau}_g = \boldsymbol{c} \times M\boldsymbol{g}.$$

Let $\boldsymbol{\tau}$ be the moment without the gravity effect. The total moment applied to the robot is given by

$$\boldsymbol{\tau}_{all} = \boldsymbol{c} \times M\boldsymbol{g} + \boldsymbol{\tau}.$$

The equation of rotational motion about the origin can be expressed as follows:

$$\dot{\mathcal{L}} = \boldsymbol{c} \times M\boldsymbol{g} + \boldsymbol{\tau}. \tag{3.37}$$

As the moment $\boldsymbol{\tau}$, we consider only **the ground reaction moment** applied on the robot. For a robot to stand still, the moment should balance with the gravitational force. If the ground reaction moment does not balance, the angular momentum rapidly increases. This is **the fall.**

3.3.2 Momentum

1 Center of Mass

No matter how complicated a robot structure has, it is ultimately a collection of atoms. Let us assume that a humanoid robot is composed of N points of

---

[5] As another examples of $\boldsymbol{f}$, we can consider the reaction force by pushing an object and the drag at the time of a wind blowing.

mass. Let m_i be the mass of the i-th point. The total mass of the robot is given by

$$M = \sum_{i=1}^{N} m_i.$$

Let $\boldsymbol{p}_i$ be the position of the i-th mass point. The position of the robot's center of mass is given by

$$\boldsymbol{c} = \sum_{i=1}^{N} m_i \boldsymbol{p}_i / M. \tag{3.38}$$

Differentiating the above equation yields

$$\dot{\boldsymbol{c}} = \sum_{i=1}^{N} m_i \dot{\boldsymbol{p}}_i / M \tag{3.39}$$

where $m_i \dot{\boldsymbol{p}}_i$ is the momentum of the i-th mass point. As a sum of the momentum of the individual point masses, the total momentum of the robot is expressed by

$$\mathcal{P} = \sum_{i=1}^{N} m_i \dot{\boldsymbol{p}}_i. \tag{3.40}$$

Here, (3.33) in the previous section can be obtained from (3.39) and (3.40).

$$\dot{\boldsymbol{c}} = \mathcal{P}/M.$$

2 Dynamics of Momentum

We will derive the dynamics of the robot's momentum. The equation of motion of the i-th point mass is given by

$$m_i \ddot{\boldsymbol{p}}_i = \sum_{j=1}^{N} \boldsymbol{f}_{ij}^{int} + \boldsymbol{f}_i^{ext} \tag{3.41}$$

where $\boldsymbol{f}_{ij}^{int}$ and $\boldsymbol{f}_i^{ext}$ denote the force applied to the i-th point mass from the j-th one and the force applied to the i-th point mass from the outside the robot. However, due to the law of action and reaction, we obtain

$$\boldsymbol{f}_{ij}^{int} = -\boldsymbol{f}_{ji}^{int} \quad (i \neq j).$$

We note that, since the force applied by the i-th object from itself is zero, we have $\boldsymbol{f}_{ii}^{int} = 0$.

Taking the above relationship into consideration and summing (3.41) for all point masses of the robot, the force applied to one point mass from another one is canceled,

$$\sum_{i=1}^{N} m_i \ddot{\boldsymbol{p}}_i = \sum_{i=1}^{N} \boldsymbol{f}_i^{ext}. \tag{3.42}$$

Equation (3.34) in the previous section can be obtained by using $\sum_{i=1}^{N} m_i \ddot{\boldsymbol{p}}_i = \dot{\mathcal{P}}$ and replacing the sum of the external forces by $\sum_{i=1}^{N} \boldsymbol{f}_i^{ext} = \boldsymbol{f}_{all}$

$$\dot{\mathcal{P}} = \boldsymbol{f}_{all}.$$

We confirmed that the robot's momentum does not depend on the internal forces but depend on the external forces. We note that, without depending on the robot's structure, this equation can be always satisfied even if the robot is composed of flexible materials or liquids.

3.3.3 Angular Momentum

1 Angular and Linear Momentum

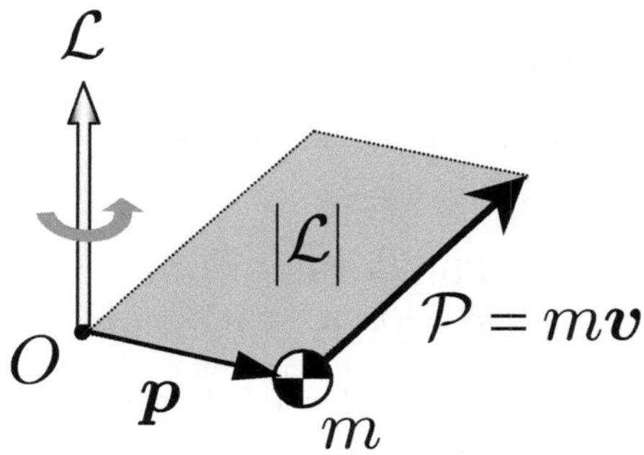

Fig. 3.18 Relationship between Linear and Angular Momentum $\mathcal{L} = \boldsymbol{p} \times \mathcal{P}$

As shown in Fig. 3.18, the angular momentum of the i-th point mass about the origin is defined by

$$\mathcal{L}_i = \boldsymbol{p}_i \times \mathcal{P}_i. \tag{3.43}$$

We should take notice for the following two remarks.

- The angular momentum is a vector and is expressed by an arrow in the three dimensional space[6].
- The angular momentum is not only a property for rotational motion. For example, by using the above equation, we can calculate the angular momentum of a point mass moving straight with constant speed. In such a case, the angular momentum keeps constant (Conservation of angular momentum).

[6] As well as the angular velocity, it is a pseudo vector.

Let us consider an arbitrary reference point expressed by a vector $\boldsymbol{r}$ other than the origin. Let $\mathcal{L}^{(r)}$ be the angular momentum about the reference point. The angular momentum of the i-th point mass about the reference point is given by

$$\mathcal{L}_i^{(r)} = (\boldsymbol{p}_i - \boldsymbol{r}) \times \mathcal{P}_i. \tag{3.44}$$

The total momentum of the robot about the reference point is

$$\begin{aligned}\mathcal{L}^{(r)} &= \sum_{i=1}^{N} (\boldsymbol{p}_i - \boldsymbol{r}) \times \mathcal{P}_i \\ &= \sum_{i=1}^{N} \boldsymbol{p}_i \times \mathcal{P}_i - \boldsymbol{r} \times \sum_{i=1}^{N} \mathcal{P}_i.\end{aligned}$$

Therefore,

$$\mathcal{L}^{(r)} = \mathcal{L} - \boldsymbol{r} \times \mathcal{P}. \tag{3.45}$$

For example, this equation can be used when we calculate the angular momentum of the robot about the center of mass.

2 Dynamics of Angular Momentum

Now we obtain the dynamics of angular momentum. Differentiating (3.43) with respect time yields

$$\begin{aligned}\dot{\mathcal{L}}_i &= \dot{\boldsymbol{p}}_i \times \mathcal{P}_i + \boldsymbol{p}_i \times \dot{\mathcal{P}}_i \\ &= \dot{\boldsymbol{p}}_i \times (m_i \dot{\boldsymbol{p}}_i) + \boldsymbol{p}_i \times m_i \ddot{\boldsymbol{p}}_i.\end{aligned}$$

Since the first term of the right-hand side is 0, we obtain

$$\dot{\mathcal{L}}_i = \boldsymbol{p}_i \times m_i \ddot{\boldsymbol{p}}_i \tag{3.46}$$

Substituting the equation of motion (3.41) in the previous section into (3.46), we obtain

$$\begin{aligned}\dot{\mathcal{L}}_i &= \boldsymbol{p}_i \times (\sum_{j=1}^{N} \boldsymbol{f}_{ij}^{int} + \boldsymbol{f}_i^{ext}) \\ &= \sum_{j=1}^{N} \boldsymbol{p}_i \times \boldsymbol{f}_{ij}^{int} + \boldsymbol{p}_i \times \boldsymbol{f}_i^{ext}.\end{aligned} \tag{3.47}$$

Since the total angular momentum is the sum of that of point masses

$$\dot{\mathcal{L}} = \sum_{i=1}^{N} \sum_{j=1}^{N} \boldsymbol{p}_i \times \boldsymbol{f}_{ij}^{int} + \sum_{i=1}^{N} \boldsymbol{p}_i \times \boldsymbol{f}_i^{ext}. \tag{3.48}$$

where the first term of the right-hand side is expressed by

$$\boldsymbol{p}_i \times \boldsymbol{f}_{ij}^{int} + \boldsymbol{p}_j \times \boldsymbol{f}_{ji}^{int} = (\boldsymbol{p}_i - \boldsymbol{p}_j) \times \boldsymbol{f}_{ij}^{int} = \boldsymbol{r}_{ij} \times \boldsymbol{f}_{ij}^{int}, \tag{3.49}$$

where $\boldsymbol{r}_{ij}$ denotes the vector from the j-th point mass to the i-th one. Generally speaking, since vectors of the action and the reaction forces between two point masses are on the line connecting them, we obtain

$$\boldsymbol{r}_{ij} \times \boldsymbol{f}_{ij}^{int} = 0.$$

The first term of the right-hand side of (3.48) becomes zero, therefore,

$$\dot{\mathcal{L}} = \sum_{i=1}^{N} \boldsymbol{p}_i \times \boldsymbol{f}_i^{ext}.$$

Since the right-hand side of the above equation expresses the moment of external forces about the origin, we obtain (3.36) in Section 3.3.1 by replacing the external forces by $\boldsymbol{\tau}_{all}$

$$\dot{\mathcal{L}} = \boldsymbol{\tau}_{all}.$$

We confirmed that the angular momentum about the origin does not depend on the internal force but depends on the moment applied from the outside the robot. Also, this equation is always satisfied regardless of the structure and the materials of the robot[7].

3.3.4 Angular Momentum and Inertia Tensor of Rigid Body

A rigid body is an ideal object which is stiff enough not to deform. A robot is usually analyzed by assuming that it is composed of some rigid bodies connected by joints[8].

We now formulate the angular momentum of a rigid body. Let us assume that a rigid body floats in the space and rotates without affected by external forces. As explained in Chapter 2, the rotational velocity of a rigid body can be expressed by the angular velocity vector $\boldsymbol{\omega}$. Let us also assume that the origin of the reference coordinate system coincides with its center of mass. The velocity at a point in the rigid body can be expressed by

$$\boldsymbol{v}_i = \boldsymbol{v}(\boldsymbol{p}_i) = \boldsymbol{\omega} \times \boldsymbol{p}_i. \tag{3.50}$$

---

[7] Derivation of the momentum and the angular momentum of the point masses in this book followed the Goldstein's Classical Mechanics [36].

[8] While this is an approximation, an analysis is accurate enough even if the robot is modeled by the sum of rigid bodies.

Substituting (3.50) into (3.43), we can calculate the total angular momentum of a rigid body by

$$\begin{aligned}\mathcal{L} &= \sum_i \boldsymbol{p}_i \times (m_i \boldsymbol{\omega} \times \boldsymbol{p}_i) \\ &= \sum_i m_i \boldsymbol{p}_i \times (-\boldsymbol{p}_i \times \boldsymbol{\omega}) \\ &= (\sum_i m_i \widehat{\boldsymbol{p}}_i \widehat{\boldsymbol{p}}_i^T)\, \boldsymbol{\omega}.\end{aligned}$$

We can see that the angular momentum of a rigid body is expressed by a angular velocity vector multiplied by a coefficient matrix. This matrix is called **inertia tensor** and is denoted by $\boldsymbol{I}$

$$\boldsymbol{I} \equiv \sum_i m_i \widehat{\boldsymbol{p}}_i \widehat{\boldsymbol{p}}_i^T. \tag{3.51}$$

As we can see from its definition, $\boldsymbol{I}$ is a 3×3 symmetric matrix. The angular momentum of a rigid body can be calculated by the angular velocity vector multiplied by the inertia tensor $\boldsymbol{I}$

$$\mathcal{L} = \boldsymbol{I}\boldsymbol{\omega}. \tag{3.52}$$

We can obtain the inertia tensor of an object with arbitrary shape and with arbitrary density distribution $\nu(\boldsymbol{p})$ by simply writing (3.51) for continuous systems

$$\boldsymbol{I} = \int_V \nu(\boldsymbol{p}) \widehat{\boldsymbol{p}} \widehat{\boldsymbol{p}}^T dV. \tag{3.53}$$

We do not need to calculate the inertia tensor of any objects with uniform density since the inertia tensor with typical shape can be found in various text books or websites. For example, for a cuboid which length of each edge is l_x, l_y and l_z and mass is m, it is

$$\boldsymbol{I} = \begin{bmatrix} \frac{m}{12}(l_y^2 + l_z^2) & 0 & 0 \\ 0 & \frac{m}{12}(l_x^2 + l_z^2) & 0 \\ 0 & 0 & \frac{m}{12}(l_x^2 + l_y^2) \end{bmatrix}. \tag{3.54}$$

Let us consider a cuboid whose length of each edge is $0.1 \times 0.4 \times 0.9$ [m^3] and mass is 36 kg. The inertia tensor of this object is given by

$$\bar{\boldsymbol{I}} = \begin{bmatrix} 2.91 & 0 & 0 \\ 0 & 2.46 & 0 \\ 0 & 0 & 0.51 \end{bmatrix} \text{ [kgm}^2\text{]}.$$

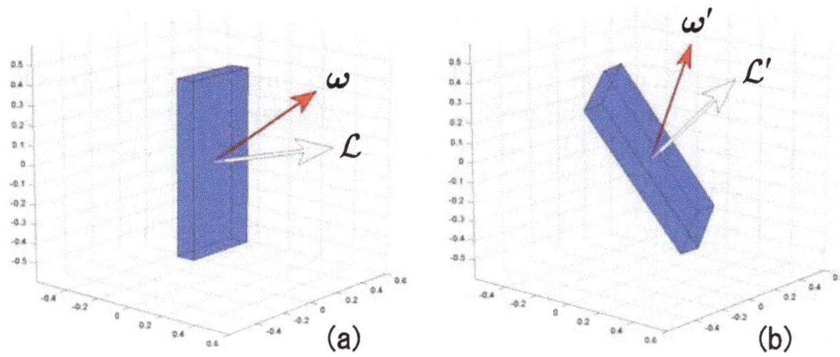

Fig. 3.19 (a)Angular Velocity Vector $\boldsymbol{\omega}$ and Angular Momentum Vector $\mathcal{L}$ of a Rigid Body (Reference Posture) (b)Rotation of Rigid Body (Relative position of both vectors $\boldsymbol{\omega}'$ and $\mathcal{L}'$ with respect to the rigid body do not change.)

Figure 3.19(a) shows the angular momentum vector of this object which angular velocity is $[1\ 1\ 1]^T$[rad/s]. We can see from this figure that the direction of the angular momentum is usually different from that of the angular velocity.

Now, Fig. 3.19(b) shows the rotation of the rigid body by multiplying a rotation matrix $\boldsymbol{R}$. If we consider that the object is rotated due to the change of view point, we can easily understand that the relative position of both the angular velocity vector and the angular momentum vector with respect to the rigid body do not change. In other words, due to the change of view point, $\boldsymbol{\omega}, \mathcal{L}$ rotates along with the rotation of the rigid body

$$\boldsymbol{\omega}' = \boldsymbol{R}\boldsymbol{\omega} \tag{3.55}$$

$$\mathcal{L}' = \boldsymbol{R}\mathcal{L}. \tag{3.56}$$

On the other hand, from (3.52), the angular momentum in the reference posture is calculated by

$$\mathcal{L} = \bar{\boldsymbol{I}}\boldsymbol{\omega}.$$

Substituting the above equation into (3.56) and replacing $\boldsymbol{\omega}$ by $\boldsymbol{\omega}$ using (3.55), we obtain

$$\mathcal{L}' = (\boldsymbol{R}\bar{\boldsymbol{I}}\boldsymbol{R}^T)\boldsymbol{\omega}'. \tag{3.57}$$

We can regard that $\boldsymbol{R}\bar{\boldsymbol{I}}\boldsymbol{R}^T$ is the new inertia tensor. Therefore, the inertia tensor of a rigid body in any attitude can be calculated from the inertia tensor $\bar{\boldsymbol{I}}$ at the reference posture and its posture $\boldsymbol{R}$ using

$$\boldsymbol{I} = \boldsymbol{R}\bar{\boldsymbol{I}}\boldsymbol{R}^T. \tag{3.58}$$

3.3.5 Calculation of Robot's Center of Mass

From the above discussion, we will calculate the physical parameters for the dynamics. We first show the method to calculate the robot's center of mass. Given the position and orientation of all links, it can be calculated by the following steps:

Step 1. Calculate the center of mass position of each link in the world coordinates.
Step 2. Calculate the sum of the moment about the origin generated by the mass of each link.
Step 3. The position of the center of mass is obtained by dividing the moment by the total mass.

Let us assume that the center of mass of the j-th link in its local coordinates is already known as $\bar{\boldsymbol{c}}_j$. The center of mass of the j-th link in the world coordinate frame is given by

$$\boldsymbol{c}_j = \boldsymbol{p}_j + \boldsymbol{R}_j \bar{\boldsymbol{c}}_j, \tag{3.59}$$

where $(\boldsymbol{p}_j, \boldsymbol{R}_j)$ denotes the position and orientation of the link. The total center of mass can be obtained by dividing the sum of the moment by the total mass

$$\boldsymbol{c} = \sum_{j=1}^{N} m_j \boldsymbol{c}_j / M. \tag{3.60}$$

The program calculating the moment about the origin of the world coordinates is shown in Fig. 3.20. By using this program, we can calculate the robot's center of mass as shown in Fig. 3.21.

```
function mc = calcMC(j)
global uLINK

if j == 0
   mc = 0;
else
   mc = uLINK(j).m * (uLINK(j).p + uLINK(j).R * uLINK(j).c );
   mc = mc + calcMC(uLINK(j).sister) + calcMC(uLINK(j).child);
end
```

Fig. 3.20 calcMC.m Calculation of the total moment about the origin of the world coordinates

```
function com = calcCoM()
global uLINK

M   = TotalMass(1);
MC  = calcMC(1);
com = MC / M;
```

Fig. 3.21 calcCoM.m Calculation of the position of center of mass

3.3.6 Calculation of Link Speed and Angular Velocity

In the following two subsections, we will calculate the total momentum and the total angular momentum of a robot. For its preparation, we need to obtain the linear and angular velocity of all the links that compose the robot.

Like we did with the forward kinematics in the previous chapter, let us compute the velocity of a link which is connected to its parent link. We will assume that the linear and angular velocity of the parent link i are already known. With the given joint speed $\dot{q}_j$, the j-th link will have the linear and angular velocity calculated by

$$\boldsymbol{v}_j = \boldsymbol{v}_i + \boldsymbol{\omega}_i \times \boldsymbol{R}_i \boldsymbol{b}_j \tag{3.61}$$

$$\boldsymbol{\omega}_j = \boldsymbol{\omega}_i + \boldsymbol{R}_i \boldsymbol{a}_j \dot{q}_j. \tag{3.62}$$

Figure 3.22 shows the Matlab code to calculate the speed and angular velocity of all links using above equation and the recursive algorithm.

```
function ForwardVelocity(j)
global uLINK

if j == 0 return; end
if j ~= 1
    i = uLINK(j).mother;
    uLINK(j).v = uLINK(i).v + cross(uLINK(i).w, uLINK(i).R * uLINK(j).b);
    uLINK(j).w = uLINK(i).w + uLINK(i).R * (uLINK(j).a * uLINK(j).dq);
end
ForwardVelocity(uLINK(j).sister);
ForwardVelocity(uLINK(j).child);
```

Fig. 3.22 ForwardVelocity.m Forward Calculation of Velocity

3.3.7 Calculation of Robot's Momentum

The momentum of a robot composed of N links is given by

$$\mathcal{P} = \sum_{j=1}^{N} m_j \dot{\boldsymbol{c}}_j \tag{3.63}$$

where $\dot{\boldsymbol{c}}_j$ denotes the velocity of the j-th link's center of mass calculated by

$$\dot{\boldsymbol{c}}_j = \boldsymbol{v}_j + \boldsymbol{\omega}_j \times (\boldsymbol{R}_j \bar{\boldsymbol{c}}_j) \tag{3.64}$$

where $(\boldsymbol{v}_j, \boldsymbol{\omega}_j)$ denotes the linear and angular velocity of the j-th link. Figure 3.23 shows the program calculating the momentum of the robot.

```
function P = calcP(j)
global uLINK

if j == 0
   P = 0;
else
   c1 = uLINK(j).R * uLINK(j).c;
   P = uLINK(j).m * (uLINK(j).v + cross(uLINK(j).w , c1));
   P = P + calcP(uLINK(j).sister) + calcP(uLINK(j).child);
end
```

Fig. 3.23 calcP.m Calculation of Robot's Momentum

3.3.8 Calculation of Robot's Angular Momentum

The angular momentum of a robot composed of N links is given by

$$\mathcal{L} = \sum_{j=1}^{N} \mathcal{L}_j \tag{3.65}$$

where $\mathcal{L}_j$ denotes the angular momentum of the j-th link with respect to the origin. It is defined by

$$\mathcal{L}_j = \boldsymbol{c}_j \times \mathcal{P}_j + \boldsymbol{R}_j \bar{\boldsymbol{I}}_j \boldsymbol{R}_j^T \omega_j \tag{3.66}$$

where $\bar{\boldsymbol{I}}_j$ denotes the inertia tensor of the j-th link with respect to the local coordinate system. By using this relationship, Fig. 3.24 shows the program calculating the robot's angular momentum about the origin.

```
function L = calcL(j)
global uLINK

if j == 0
   L = 0;
else
   c1 = uLINK(j).R * uLINK(j).c;
   c  = uLINK(j).p + c1;
   P = uLINK(j).m * (uLINK(j).v + cross(uLINK(j).w , c1));
   L = cross(c, P) + uLINK(j).R * uLINK(j).I * uLINK(j).R' * uLINK(j).w;
   L = L + calcL(uLINK(j).sister) + calcL(uLINK(j).child);
end
```

Fig. 3.24 calcL.m Calculation of Robot's Angular Momentum about the Origin

3.4 Calculation of ZMP from Robot's Motion

By using the above theories and algorithms of robot's dynamics, we can calculate the ZMP for given motion of the robot.

3.4.1 Derivation of ZMP

Let us review that the ground reaction force can be expressed by using the ZMP ($\boldsymbol{p}$), the force ($\boldsymbol{f}$) and the moment ($\boldsymbol{\tau}_p$) about the vertical line including the ZMP(Section 3.1.3). Calculating the moment of the ground reaction force, we obtain

$$\boldsymbol{\tau} = \boldsymbol{p} \times \boldsymbol{f} + \boldsymbol{\tau}_p. \tag{3.67}$$

The relationships between the ground reaction force and the momentum and between the ground reaction moment and the angular momentum (Section 3.3.1) are given by

$$\dot{\mathcal{P}} = M\boldsymbol{g} + \boldsymbol{f} \tag{3.68}$$

$$\dot{\mathcal{L}} = \boldsymbol{c} \times M\boldsymbol{g} + \boldsymbol{\tau}. \tag{3.69}$$

Substituting (3.67) and (3.68) into (3.69) and solving with respect to $\boldsymbol{\tau}_p$, we obtain

$$\boldsymbol{\tau}_p = \dot{\mathcal{L}} - \boldsymbol{c} \times M\boldsymbol{g} + (\dot{\mathcal{P}} - M\boldsymbol{g}) \times \boldsymbol{p}. \tag{3.70}$$

In detail, the first and second rows of this equation are

$$\tau_{px} = \dot{\mathcal{L}}_x + Mgy + \dot{\mathcal{P}}_y p_z - (\dot{\mathcal{P}}_z + Mg)p_y = 0 \tag{3.71}$$

$$\tau_{py} = \dot{\mathcal{L}}_y - Mgx - \dot{\mathcal{P}}_x p_z + (\dot{\mathcal{P}}_z + Mg)p_x = 0 \tag{3.72}$$

where

$$\mathcal{P} = [\mathcal{P}_x \ \mathcal{P}_y \ \mathcal{P}_z]^T$$
$$\mathcal{L} = [\mathcal{L}_x \ \mathcal{L}_y \ \mathcal{L}_z]^T$$
$$\boldsymbol{c} = [x \ y \ z]^T$$
$$\boldsymbol{g} = [0 \ 0 \ -g]^T.$$

Here, we used the fact that the x and y components of the moment about the ZMP are zero.

Solving the above equations with respect to p_x and p_y, we obtain the ZMP as

$$p_x = \frac{Mgx + p_z\dot{\mathcal{P}}_x - \dot{\mathcal{L}}_y}{Mg + \dot{\mathcal{P}}_z} \tag{3.73}$$

$$p_y = \frac{Mgy + p_z\dot{\mathcal{P}}_y + \dot{\mathcal{L}}_x}{Mg + \dot{\mathcal{P}}_z} \tag{3.74}$$

where p_z denotes the height of the ground. For walking on the flat ground, we have $p_z = 0$.

For example, when a robot stands still, we have $\dot{\mathcal{P}} = \dot{\mathcal{L}} = 0$ and

$$p_x = x \tag{3.75}$$

$$p_y = y. \tag{3.76}$$

The ZMP coincides with the ground projection of the center of mass.

Figure 3.25 shows the program calculating the ZMP by using (3.73) and (3.74).

```
function [px,py] = calcZMP(c,dP,dL,pz)
global M G

px = (M*G*c(1) + pz * dP(1) - dL(2))/(M*G + dP(3));
py = (M*G*c(2) + pz * dP(2) + dL(1))/(M*G + dP(3));
```

Fig. 3.25 calcZMP.m Calculation of ZMP

Here, dP(= $\dot{\mathcal{P}}$), dL(= $\dot{\mathcal{L}}$) can be calculated by using the numerical differentiation of the momentum and the angular momentum.

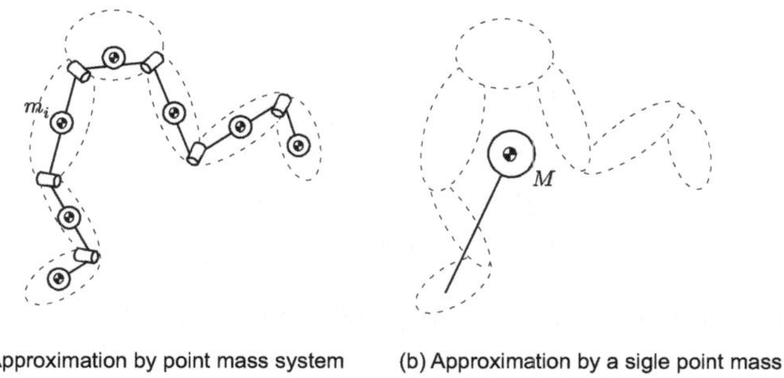

Fig. 3.26 Approximated multibody system

3.4.2 Calculation of ZMP Using Approximation

We introduce the method to calculate ZMP using the simplified models. Fig. 3.26(a) models the robot neglecting the effect of the inertia tensor of each link about its center of mass and assuming the robot as a sum of point masses. In this case, the angular momentum about the origin is given by

$$\mathcal{L} = \sum_{i=1}^{N} \boldsymbol{c}_i \times \mathcal{P}_i. \tag{3.77}$$

Substituting (3.77) into (3.73) and (3.74), the ZMP can be expressed as

$$p_x = \frac{\sum_{i=1}^{N} m_i\{(\ddot{z}_i + g)x_i - (z_i - p_z)\ddot{x}_i\}}{\sum_{i=1}^{N} m_i(\ddot{z}_i + g)} \tag{3.78}$$

$$p_y = \frac{\sum_{i=1}^{N} m_i\{(\ddot{z}_i + g)y_i - (z_i - p_z)\ddot{y}_i\}}{\sum_{i=1}^{N} m_i(\ddot{z}_i + g)} \tag{3.79}$$

where $\boldsymbol{c}_i = [x_i\ y_i\ z_i]^T$. While this equation is an approximation, the ZMP can be calculated with enough accuracy if we model each link by using multiple point masses [6].

Next, in the model shown in Fig. 3.26(b), the whole robot is modeled by a point mass. In this case, the momentum and the angular momentum about the origin are given by

$$\mathcal{P} = M\dot{\boldsymbol{c}} \tag{3.80}$$

$$\mathcal{L} = \boldsymbol{c} \times M\dot{\boldsymbol{c}}, \tag{3.81}$$

where their elements are

$$\begin{bmatrix} \dot{\mathcal{P}}_x \\ \dot{\mathcal{P}}_y \\ \dot{\mathcal{P}}_z \end{bmatrix} = \begin{bmatrix} M\ddot{x} \\ M\ddot{y} \\ M\ddot{z} \end{bmatrix} \tag{3.82}$$

$$\begin{bmatrix} \dot{\mathcal{L}}_x \\ \dot{\mathcal{L}}_y \\ \dot{\mathcal{L}}_z \end{bmatrix} = \begin{bmatrix} M(y\ddot{z} - z\ddot{y}) \\ M(z\ddot{x} - x\ddot{z}) \\ M(x\ddot{y} - y\ddot{x}) \end{bmatrix}. \tag{3.83}$$

Substituting the above equation to (3.73) and (3.74), the ZMP can be given by

$$p_x = x - \frac{(z - p_z)\ddot{x}}{\ddot{z} + g} \tag{3.84}$$

$$p_y = y - \frac{(z - p_z)\ddot{y}}{\ddot{z} + g}. \tag{3.85}$$

We will use (3.84) in Chapter 4 to generate the biped gait pattern.

3.5 Some Notes for ZMP

3.5.1 Two Explanations

By intuitively explaining the relationship between the robot's motion and the ZMP, figures of the point-mass model like Fig. 3.27(a) is often introduced. Here, $-M\ddot{x}$ denotes the virtual force called the **inertial force** expressing the reaction generated by the acceleration of an object[8]. Fig. 3.27(a) shows that the inertial force and the gravity force balance with the ground reaction force. Here, the forces acting on the robot is apparently the gravity and the ground reaction forces as shown in 3.27(b). Since it is difficult to show the balance of forces, we introduced the inertial force. However, we can explain without introducing the inertial force. In Fig.3.27(c), opposing the gravity force, the center of mass goes up due to the effect of the ground reaction force. At the same time, the center of mass is accelerated due to the effect of it [23]. In this case, the ground reaction force is decomposed into the gravity compensation and the acceleration forces. Of course, the right results can be expected by introducing both explanations.

3.5.2 Does ZMP Exist Outside the Support Polygon due to the Acceleration of the Center of Mass?

The discussions like "Depending on the motion of the robot, may the ZMP leave the support polygon?" often occurs. Of course, the conclusion is that "the ZMP never exists outside the support polygon" [37, 88].

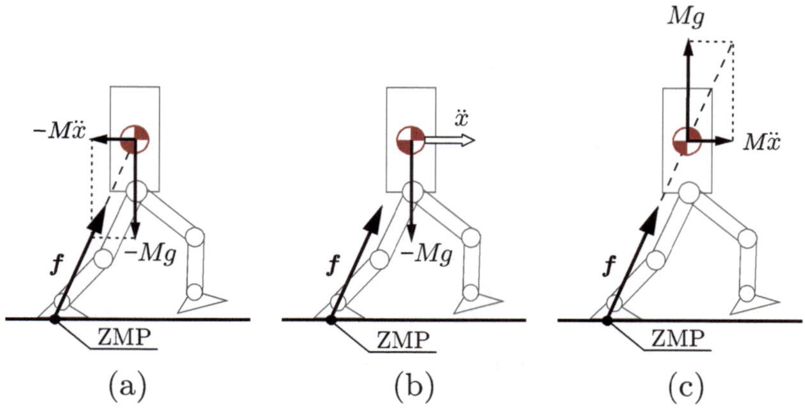

Fig. 3.27 Relationship between Robot's Motion and ZMP (a) Explanation using Inertial Force (b) Force acting on Robot (c) Explanation using the Gravity Compensation Force and Acceleration Force

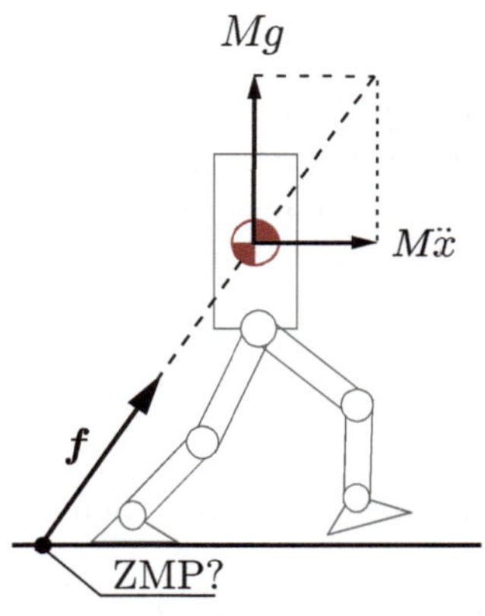

Fig. 3.28 May the ZMP leave the support polygon if the robot strongly accelerates?

However, as shown in Fig. 3.28, what will happen when robot modeled by a point mass moves horizontally with high acceleration? If there is enough friction between the sole and the ground, the horizontal acceleration will not be barred. As explained above, the ZMP exists on the line defined by the gravity and inertial forces, and their values can be obtained by substituting $\ddot{z} = 0$ and $p_z = 0$ into (3.84)

$$p_x = x - \frac{z\ddot{x}}{g}.$$

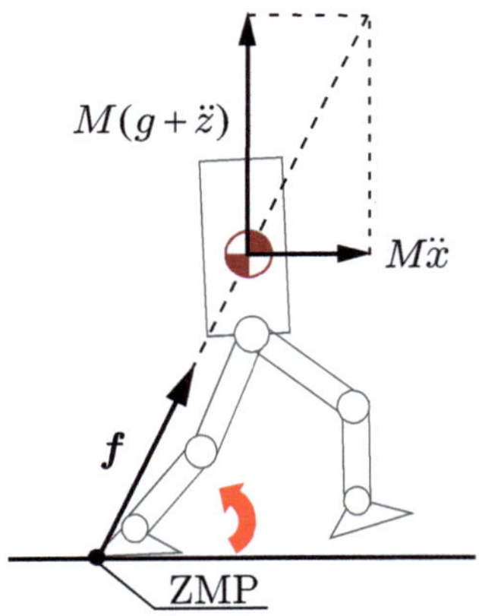

Fig. 3.29 The robot begins to rotate about the heel since the ZMP exists at the heel due to the highly accelerated center of mass. Since the acceleration in the vertical direction is generated, the ZMP remains in the support polygon.

The larger the acceleration of the robot becomes, the further from the support polygon the ZMP can be!

Using Fig. 3.29, we will explain the mistake in this discussion. Since the center of mass is highly accelerated, the ZMP is shifted to the heel. Then the robot will begin to rotate about the heel. Since the acceleration in the vertical direction is generated, we have $\ddot{z} > 0$. Taking this effect into consideration, the ZMP should be calculated by

$$p_x = x - \frac{z\ddot{x}}{\ddot{z} + g}.$$

Since $\ddot{z}$ will increase according to the increment of $\ddot{x}$, the ZMP remains in the support polygon[9].

More concretely, to calculate the ZMP for given motion of the robot using (3.73) and (3.74), we need to introduce either one of the following preconditions

Precondition A
: The sole keeps the surface contact with the ground since it is fixed to the ground.

Precondition B
: The posture and the absolute linear and angular velocity of the robot can be measured.

We usually use the **Precondition A** when calculating the ZMP from the simulated motion of the robot. In this case, the ZMP obtained from (3.73)

[9] This is the principle of the "Model ZMP Control" proposed by Honda Motor Co., Inc. For more concrete discussion, see p.152.

and (3.74) may exist outside the support polygon[10]. However, to realize this situation by real robots, the sole should be fixed to the ground by using suction cups or magnets. Ordinary humanoid robots cannot realize this situation since the sole will break contact with the ground.

On the other hand, to calculate the ZMP of a real robot by using (3.73) and (3.74), we need to introduce the **precondition B**. This ZMP coincides with the ZMP measured from the force/torque sensors at the foot and never exists outside the support polygon.

3.5.3 Limitation of ZMP

Since the ZMP is equivalent to the center of pressure, its physical meaning is very clear. Also, the relationship between the ZMP and the linear/angular momentum of the robot can be expressed by a simple equation. Thus, the ZMP can be a powerful tool to plan the walking motion on the flat ground with enough friction.

On the other hand, the ZMP cannot be used for the following cases:

[A] We want to determine whether or not the sole slips on the ground surface.
[B] The ground surface is not flat.
[C] The arms or the hands of a humanoid robot contact the environment.

As for [A], only by using the ZMP information, we cannot judge the slip on the ground as described in Section 3.1. On the other hand, as for [B] and [C], although the position of the ZMP changes according to the friction force, there is a case where the friction force cannot be determined uniquely for given motion of the robot. Also, the internal forces among the contact points do not affect the position of the ZMP.

The above problems arise since the ZMP is the two dimensional information of the ground reaction force while six dimensional information of the force/moment is required to determine the transition of contact states[11].

Here, we introduce an approach taking the full six dimensional force/moment into consideration [39].

Let us assume the robot to be a single rigid body. At the point in the rigid body specified by the vector $\boldsymbol{p}_B$, the force/moment applied by the robot to the ground can be obtained by

$$\boldsymbol{f}_B^G = M(\boldsymbol{g} - \ddot{\boldsymbol{c}}) \tag{3.86}$$

$$\boldsymbol{\tau}_B^G = -\dot{\mathcal{L}}^{(c)} + M(\boldsymbol{c} - \boldsymbol{p}_B) \times (\boldsymbol{g} - \ddot{\boldsymbol{c}}) \tag{3.87}$$

---

[10] This ZMP is called the **IZMP(Imaginary ZMP)** [89].

[11] As for [B] and [C], the approach focusing on the moment about the edges of the convex polygon is proposed [80]. However, the problem remains in the treatment of the friction forces.

where $\mathcal{L}^{(c)}$ denotes the angular momentum about the center of mass. Let us consider the infinitesimal translational/rotational displacement $[(\delta \boldsymbol{p}_B)^T \ (\boldsymbol{\omega}_B)^T]^T$ of the rigid body. Let $S_1(\delta \boldsymbol{p}_B, \boldsymbol{\omega}_B)$ be the set of the translational/rotational displacement of the robot without interfering the environment. In this case, if the following equation is satisfied

$$\forall(\delta \boldsymbol{p}_B, \boldsymbol{\omega}_B) \in S_1(\delta \boldsymbol{p}_B, \boldsymbol{\omega}_B)$$
$$\begin{bmatrix}(\boldsymbol{f}_B^G)^T & (\boldsymbol{\tau}_B^G)^T\end{bmatrix} \begin{bmatrix}\delta \boldsymbol{p}_B \\ \boldsymbol{\omega}_B\end{bmatrix} \leq 0 \tag{3.88}$$

the robot will not move since the work done by $(\boldsymbol{f}_B^G, \ \boldsymbol{\tau}_B^G)$ is not positive. In other words, the contact state between the robot and the environment will not change. In this method, the change of the contact state is determined by checking whether or not $\boldsymbol{f}_B^G$ and $\boldsymbol{\tau}_B^G$ satisfy the above inequalities for given motion of the robot. This method gives the necessary and sufficient condition to keep a contact state for the robot walking on the flat ground with enough friction. On the other hand, this becomes a necessary condition to keep a contact state when the friction coefficient between the robot and the ground is low and when the contact points are not limited to the flat ground.

3.6 Appendix: Convex Set and Convex Hull

We explain the convex set and the convex hull which was introduced in Section 3.1.1 to define the support polygon and in Section 3.1.3 to show the region of the ZMP. In the robotics research community, the convex hull is used for grasp analysis by robotic hands and collision avoidance between objects. Also, in the research community of mathematical programming, the convex set and the convex hull are the important basic concepts. For more concrete discussions, please refer [104] for example[12].

Convex Set:
A subset S of R^n is defined to be the convex set if

$$\alpha \boldsymbol{p}_1 + (1-\alpha)\boldsymbol{p}_2 \in S \tag{3.89}$$

is satisfied for any $\boldsymbol{p}_1, \boldsymbol{p}_2 \in S$ and α, $0 \leq \alpha \leq 1$. As shown in Fig. 3.30, (3.89) expresses the segment between $\boldsymbol{p}_1$ and $\boldsymbol{p}_2$ when S is a subset of R^2. In other words, if a segment formed by connecting arbitrary two points included in the set S is also included in S, this set is defined to be the convex set.

[12] For example, the convex set and the convex hull can be defined for sets which are not bounded. However, since the contact area between the sole and the ground is bounded, we explain only for the bounded set in this book.

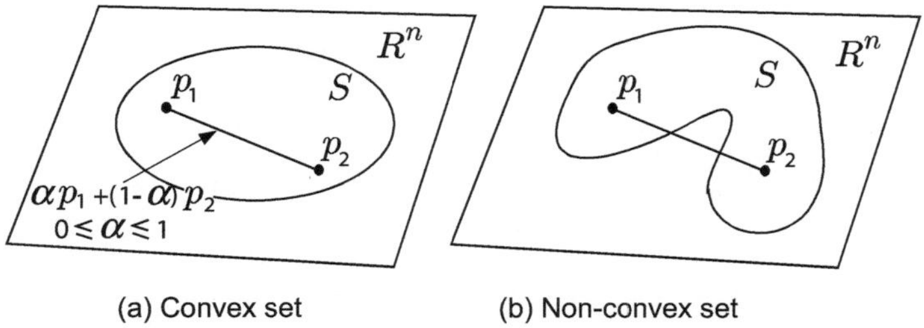

Fig. 3.30 Definition of Convex Set

Convex Hull:

For a subset S of R^n, the minimum convex set including S is defined to be the convex hull co$S$. The set S shown in Fig. 3.31(a) is not the convex set. The minimum region formed by enclosing the set S by rope corresponds to the convex hull.

Let us consider the case where the convex hull is the bounded convex polyhedron. Let $\boldsymbol{p}_j$ $(j = 1, \cdots, N)$ be the vector of the edges. The convex hull is defined by

$$\mathrm{co}S = \left\{ \sum_{j=1}^{N} \alpha_j \boldsymbol{p}_j \,\middle|\, \alpha_j \geq 0,\ \sum_{j=1}^{N} \alpha_j = 1,\ \boldsymbol{p}_j \in S\ (j = 1, \cdots, N) \right\}. \quad (3.90)$$

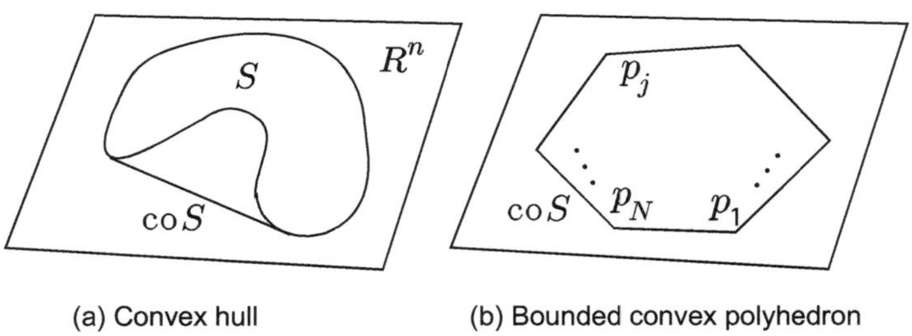

Fig. 3.31 Definition of Convex Hull

Convex Shell

Chapter 4
Biped Walking

What is the meaning of the word "walk"? Oxford Advanced Learner's Dictionary gives us a concise definition.

> **walk:**
> move along at a moderate pace by lifting up and putting down each foot in turn, so that one foot is on the ground while the other is being lifted
> (Oxford Advanced Learner's Dictionary, Oxford University Press)

Therefore, at least one foot must be in contact with the ground at any moment during walking. There exist two kind of walking, namely, static walking and dynamic walking. In "static walking", the projection of the center of mass never leaves the support polygon during the walking. In "dynamic walking", there exist periods when the projection of the center of mass leaves the support polygon.

$$\text{Walk} \begin{cases} \text{static walk} \\ \text{dynamic walk} \end{cases}$$

Most toy robots perform static walking using large feet. This is not interesting from the view point of control engineering since it is fairly easy. Nevertheless, human feet are too small with respect to the height of center of mass to perform static walking. Indeed we are performing dynamic walking in our daily life. The walking style with which we are so accustomed can be realized by dexterous control of the whole body balance which is essentially unstable. A biped walking machine is, therefore, beyond the scope of conventional mechanical engineering. This is the reason that so many researchers and engineers are attracted to biped walking machines as well as humanoid robots.

4.1 How to Realize Biped Walking?

Figure 4.1 shows the basic framework of biped walking control that will be used throughout this chapter. A set of time series of joint angles for desired walking is called a *walking pattern*, and to create it, we use a *walking pattern*

S. Kajita et al., *Introduction to Humanoid Robotics*, Springer Tracts in Advanced Robotics 101,
DOI: 10.1007/978-3-642-54536-8_4,

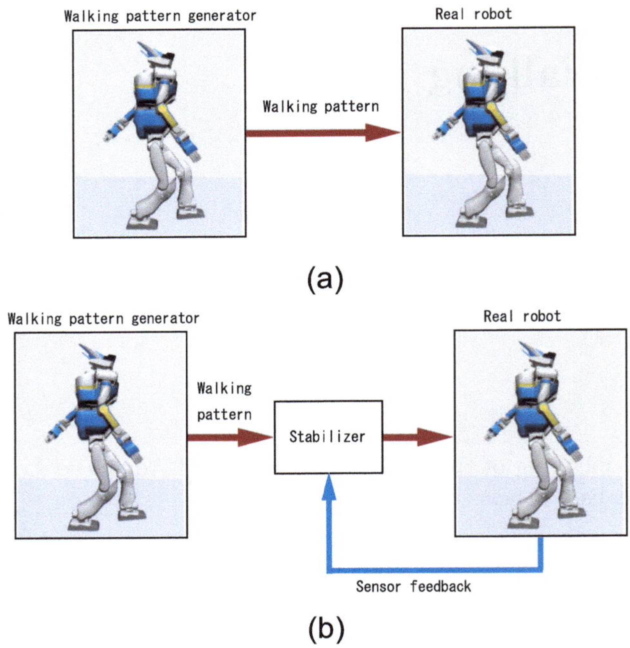

Fig. 4.1 Basic control framework in this book: (a)In an ideal condition, a biped robot can walk by following a walking pattern. (b)In a real environment, a biped robot need a stabilizer.

generator. In an ideal situation, biped walking can be realized just by giving a walking pattern to an actual robot (Fig. 4.1(a)). For this purpose, we must prepare an accurate model of the robot, a stiff mechanism which moves exactly as commanded and a perfect horizontal floor (a huge surface plate).

On the other hand, in a real situation, a life size humanoid robot can easily fall down by floor unevenness of only a few millimeters. A humanoid proportion and mass distribution tends to quickly amplify the posture error to an unstable level. To suppress this, we need the second software, which modifies the walking pattern by using gyros, accelerometers, force sensors and other devices. This is called a *stabilizer* (Fig. 4.1(b)).

The rest of this chapter is organized as follows. In section 4.2, 4.3 and 4.4, we explain walking pattern generators and explain stabilizers in section 4.5. In section 4.6, we spotlight the history of biped walking robot research. In section 4.7, we introduce a variety of biped control methods which are not restricted in the framework of Fig. 4.1.

4.2 Two Dimensional Walking Pattern Generation

In this section, we consider a basic principle of biped walking in 2D and derive an algorithm for walking pattern generation.

4.2.1 Two Dimensional Inverted Pendulum

Coarse-graining is a powerful method to handle a complex system. In celestial mechanics, researchers approximate the Sun and the planets as point masses while they have their own internal structure, and they can still calculate the orbits of the solar system with sufficient accuracy. In thermodynamics, vast states of molecules, number of the order of 10^{23} are coarse-grained as a few parameters like temperature or entropy, and it makes it possible for us to predict the thermodynamic phenomenon.

Likewise, to extract an *essence* of biped locomotion, we make three assumptions as coarse-graining on a humanoid robot with more than 30 DOF made of over thousands of mechanical and electrical parts. First, we assume that all the mass of the robot is concentrated at its center of mass (CoM). Second, we assume that the robot has massless legs, whose tips contact the ground at single rotating joints. At last, we only consider the forward/backward and the up/down motions of the robot, neglecting lateral motion. In other words, we assume the robot motion is constrained to the *sagittal plane* defined by the axis of walking direction and vertical axis. With these assumptions, we model a robot as a 2D inverted pendulum as shown in Fig. 4.2.

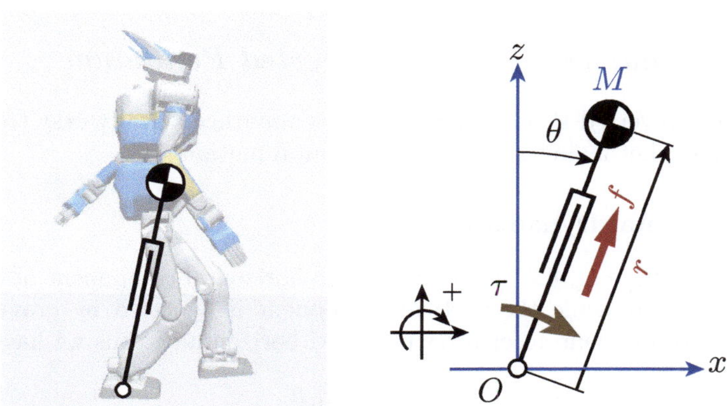

Fig. 4.2 2D inverted pendulum: The simplest model for a human or a walking robot. It consists of the center of mass (CoM) and massless telescopic leg. We measure the inclination of the pendulum θ from vertical, positive for counter clockwise rotation.

The inputs of the pendulum are the torque τ at the pivot and the kick force f at the prismatic joint along the leg. The dynamics of the pendulum are described as a couple of differential equations[1] as follows

$$r^2\ddot{\theta} + 2r\dot{r}\dot{\theta} - gr\sin\theta = \tau/M$$
$$\ddot{r} - r\dot{\theta}^2 + g\cos\theta = f/M.$$

We can simulate the behavior of the inverted pendulum by integrating them numerically with given input torque.

One of the important limitations is we cannot use big torque τ since the feet of biped robot is very small. If a walking robot has a point contact like a stilt, we must use

$$\tau = 0. \tag{4.1}$$

In this case, the pendulum will almost always fall down, unless the CoM is located precisely above the pivot. Even with such a simple pendulum, we have a variety of falling patterns corresponding to different kick forces, f, as illustrated in Fig. 4.3.

The most interesting case is Fig. 4.3(d) where the CoM moves horizontally under the kick force

$$f = \frac{Mg}{\cos\theta}. \tag{4.2}$$

Figure 4.4 illustrates the reason for the horizontal motion of the CoM.

Intuitively, we can say the pendulum is keeping the CoM height by extending its leg as fast as it is falling. We call this the Linear Inverted Pendulum [115][2].

4.2.2 Behavior of Linear Inverted Pendulum

The Linear Inverted Pendulum provides us a mathematically easy treatment of dynamics. Let us investigate the horizontal motion.

1 Horizontal Dynamics

By investigating Fig. 4.4 again, we see the horizontal component of the kick force f remains while the vertical component is canceled by gravity. The horizontal component accelerates the CoM horizontally, thus we have

$$M\ddot{x} = f\sin\theta. \tag{4.3}$$

[1] These equations can be derived by using Lagrange's method. Joseph-Louis Lagrange (1736-1813) was a French mathematician born in Italy. He is famous as a discoverer of the *Lagrange points*, the most suitable area to place space colonies.

[2] This is the simplest version of Linear Inverted Pendulum (LIP). The concept will be extended throughout of this chapter. Also, the LIP is an example of Zero-dynamics which is discussed in the textbook by Westervelt et al. [24]

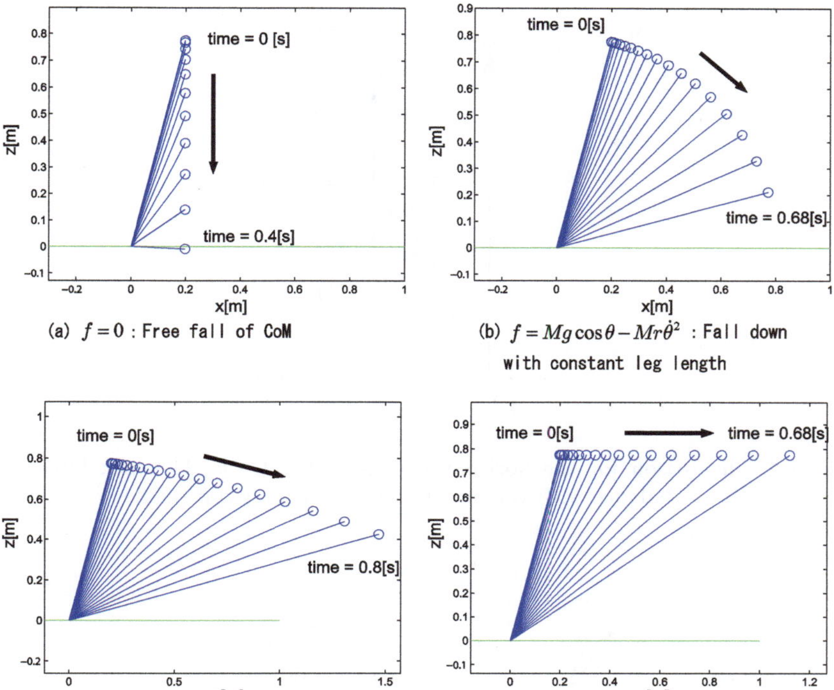

(a) $f = 0$: Free fall of CoM

(b) $f = Mg\cos\theta - Mr\dot{\theta}^2$: Fall down with constant leg length

(c) $f = Mg$: Fall down and acceleration

(d) $f = Mg/\cos\theta$: CoM accelerates while keeping the initial height

Fig. 4.3 Falling inverted pendulum under the various kick force f. The pivot torque is kept zero ($\tau = 0$) at all time.

By substituting (4.2), we get

$$M\ddot{x} = \frac{Mg}{\cos\theta}\sin\theta = Mg\tan\theta = Mg\frac{x}{z}$$

where, x, z gives the CoM of the inverted pendulum. By rewriting above equation, we obtain a differential equation for the horizontal dynamics of the CoM

$$\ddot{x} = \frac{g}{z}x. \tag{4.4}$$

Since we have constant z in a Linear Inverted Pendulum, we can easily solve this ordinary differential equation

$$x(t) = x(0)\cosh(t/T_c) + T_c\dot{x}(0)\sinh(t/T_c) \tag{4.5}$$

$$\dot{x}(t) = x(0)/T_c\sinh(t/T_c) + \dot{x}(0)\cosh(t/T_c) \tag{4.6}$$

$$T_c \equiv \sqrt{z/g}$$

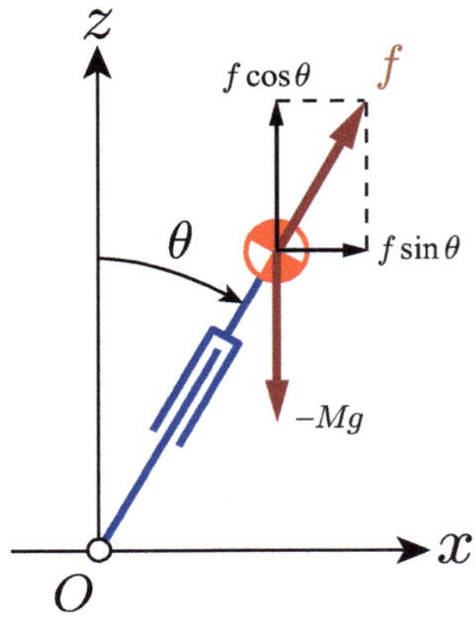

Fig. 4.4 The reason for the horizontal locus of the CoM. The kick force $f = Mg/\cos\theta$ always balance with the gravity acting on the point mass.

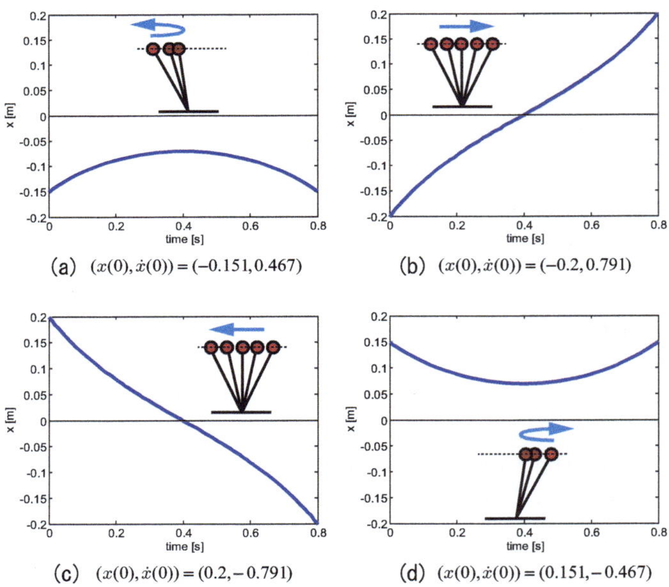

Fig. 4.5 Linear Inverted Pendulum under various initial conditions. CoM height: $z = 0.8$ m.

where T_c is the time constant depending the height of the CoM and gravity acceleration.

The initial position and velocity are given by $x(0), \dot{x}(0)$, which together are called the *initial conditions*. Figure 4.5 shows the motions under various initial conditions.

2 Time for Transfer

In many cases, we want to know the time that CoM takes to travel from one point to another. When an initial condition $(x_0, \dot{x}_0)$ and a condition $(x_1, \dot{x}_1)$ at certain moment is given, they are connected by following equations by using (4.5) and (4.6)

$$x_1 = x_0 \cosh(\tau/T_c) + T_c\dot{x}_0 \sinh(\tau/T_c) \tag{4.7}$$

$$\dot{x}_1 = x_0/T_c \sinh(\tau/T_c) + \dot{x}_0 \cosh(\tau/T_c) \tag{4.8}$$

where τ is the period that the CoM takes a trip from $(x_0, \dot{x}_0)$ to $(x_1, \dot{x}_1)$.

By using the relationship of $\cosh(x) = \frac{e^x+e^{-x}}{2}$, $\sinh(x) = \frac{e^x-e^{-x}}{2}$, these equations can be rewritten as

$$x_1 = \frac{x_0 + T_c\dot{x}_0}{2} e^{\tau/T_c} + \frac{x_0 - T_c\dot{x}_0}{2} e^{-\tau/T_c}, \tag{4.9}$$

$$\dot{x}_1 = \frac{x_0 + T_c\dot{x}_0}{2T_c} e^{\tau/T_c} - \frac{x_0 - T_c\dot{x}_0}{2T_c} e^{-\tau/T_c}. \tag{4.10}$$

From (4.9)+$T_c\times$(4.10), we get

$$x_1 + T_c\dot{x}_1 = (x_0 + T_c\dot{x}_0)e^{\tau/T_c}.$$

Therefore, τ can be calculated as

$$\tau = T_c \ln \frac{x_1 + T_c\dot{x}_1}{x_0 + T_c\dot{x}_0}. \tag{4.11}$$

Similarly, we can calculate τ from (4.9)$-T_c\times$(4.10) as

$$\tau = T_c \ln \frac{x_0 - T_c\dot{x}_0}{x_1 - T_c\dot{x}_1}. \tag{4.12}$$

Basically, (4.11) and (4.12) gives exactly the same result, except in the singular case where both of the numerator and denominator go close to zero in one of the equations.

4.2.3 Orbital Energy

To understand motions of a linear inverted pendulum intuitively, it is useful to imagine a potential as shown in Fig. 4.6. Figure 4.6(a) shows the case which the CoM changes its moving direction when the initial velocity is not enough. In Fig. 4.6(b), the CoM keeps original moving direction due to sufficient initial speed. In this case, the CoM undergoes minimum speed at the top of the potential where the CoM is just above the ankle pivot.

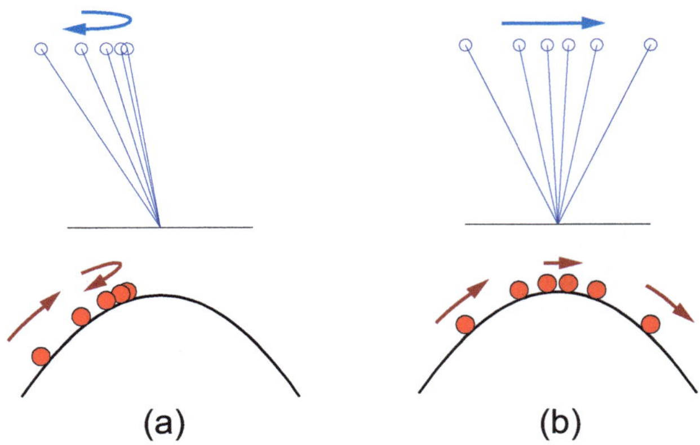

Fig. 4.6 Linear Inverted Pendulum and imaginary potential

Let us calculate the relationship between this imaginary potential and the motion of the CoM. We multiply $\dot{x}$ on both side of the equation of motion (4.4), then integrate it

$$\dot{x}(\ddot{x} - \frac{g}{z}x) = 0$$

$$\int \{\ddot{x}\dot{x} - \frac{g}{z}x\dot{x}\}dt = constant.$$

The result is

$$\frac{1}{2}\dot{x}^2 - \frac{g}{2z}x^2 = constant \equiv E. \tag{4.13}$$

The first term of the left hand side is kinetic energy and the second term is the imaginary potential energy which is illustrated by Fig. 4.6. In this calculation, we assume energies per unit mass so that we can omit mass of the CoM.

Let us call the sum of the kinetic energy and the imaginary potential energy as, E, the *orbital energy*[3] [36]. Equation (4.13) shows that **the orbital energy is conserved in the motion of Linear Inverted Pendulum**.

[3] In mechanics, this value is called constant of motion [36].

In the case of Fig. 4.6(a), where the CoM reverts without getting to the top of the potential, the orbital energy is given by

$$E = -\frac{g}{2z}x_{rev}^2, \tag{4.14}$$

where x_{rev} is the horizontal position of the CoM at the moment the reversion. In this case the orbital energy is negative or zero.

In the case of Fig. 4.6(b), where the CoM passes the top of the potential, the orbital energy is given by

$$E = \frac{1}{2}\dot{x}_{top}^2, \tag{4.15}$$

where $\dot{x}_{top}(> 0)$ is the speed at the instant that the CoM passes above of the ankle pivot. In this case, the orbital energy is positive.

Suppose we have obtained the CoM position and the speed of an inverted pendulum at certain instant. By checking the sign of the orbital energy E calculated by (4.13), we can immediately predict whether the CoM will pass the potential or not. If $E > 0$, we can predict the CoM speed at the top of the potential as

$$|\dot{x}_{top}| = \sqrt{2E}.$$

If $E < 0$, we can predict the position where the CoM will revert as

$$|x_{rev}| = \sqrt{-\frac{2zE}{g}}.$$

4.2.4 Support Leg Exchange

Although the motion of a linear inverted pendulum is determined only by its initial condition, we can control its speed because an initial condition can be modified by the touchdown timing. As illustrated in Fig. 4.7, a support exchange of quickened touchdown will decelerate the walking speed, and a support exchange of delayed touchdown will accelerate the walking. This corresponds to our experience, for example, we put down our leg quickly on the ground for sudden stops.

Let us figure out the relationship between a support exchange and the pendulum motions. Figure 4.8 shows the moment of a support exchange. The step length is s, the position of the CoM with respect to the former contact point is x_f, and the speed of the CoM at the instant of the exchange is v_f. For simplicity, we assume that the leg support is exchanged instantly, therefore, v_f is the final speed of the previous support phase as well as the initial speed of the new support phase.

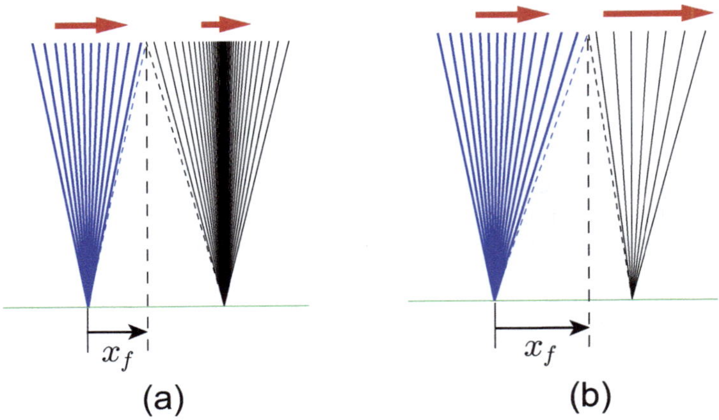

Fig. 4.7 Control of walking speed (with fixed step length) (a) Earlier touchdown of the next step results slow down (b) Later touchdown of the next step results speed up

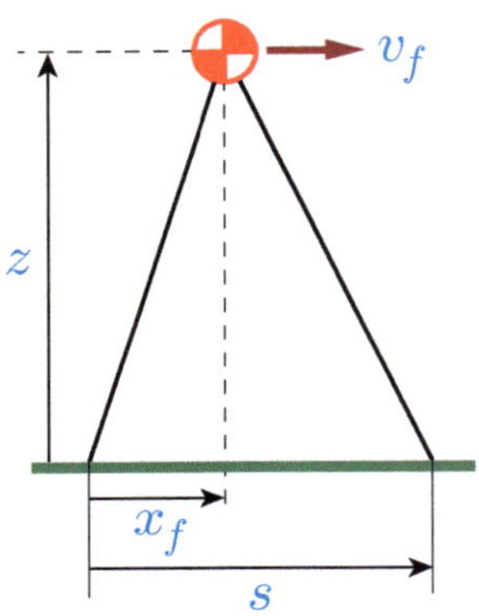

Fig. 4.8 State of a support leg exchange

By defining the orbital energies before and after the exchange as E_1 and E_2 respectively, we have

$$E_1 = -\frac{g}{2z}x_f^2 + \frac{1}{2}v_f^2 \tag{4.16}$$

$$E_2 = -\frac{g}{2z}(x_f - s)^2 + \frac{1}{2}v_f^2. \tag{4.17}$$

When the orbital energies E_1, E_2 were given, we can calculate the necessary exchange condition by eliminating v_f from (4.16), (4.17) to obtain

$$x_f = \frac{z}{gs}(E_2 - E_1) + \frac{s}{2}. \tag{4.18}$$

The speed at the moment of the exchange can be calculated from (4.16)

$$v_f = \sqrt{2E_1 + \frac{g}{z}x_f^2}. \tag{4.19}$$

4.2.5 Planning a Simple Biped Gait

Let us design a simple walking pattern using the result of former sections. We assume an ideal biped robot on a level plane walks just one step and stops (Fig. 4.9). The robot exchanges its support twice and three orbital energies must be specified.

For the walk start(a→b), the orbital energy is specified by the initial position of the CoM,

$$E_0 = -\frac{g}{2z}x_s^2.$$

For the step (b→c→d), the orbital energy is specified by the speed v_1 at the moment that CoM passes over the supporting point

$$E_1 = \frac{1}{2}v_1^2.$$

For the walk finish (d→e), the orbital energy is specified by the final position of the CoM.

$$E_2 = -\frac{g}{2z}x_e^2.$$

Using (4.18), the first support exchange condition x_{f0} is obtained from E_0, E_1, and the second support exchange condition x_{f1} is obtained from E_1, E_2. Desired walking motion is realized by controlling the swing leg so that those exchanges occur at the right time.

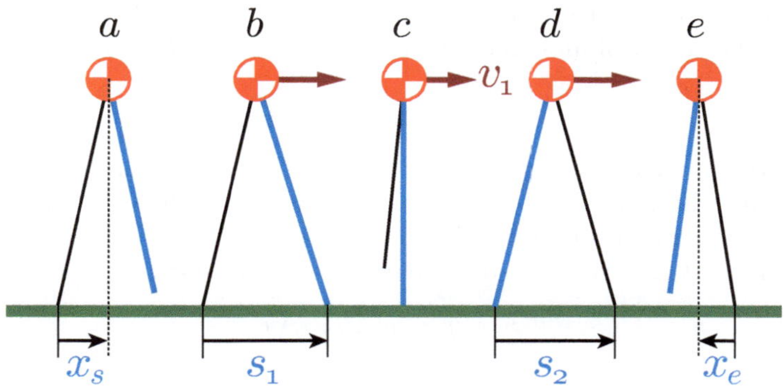

Fig. 4.9 Specification for a walk of one step forward. We need support leg exchanges twice.

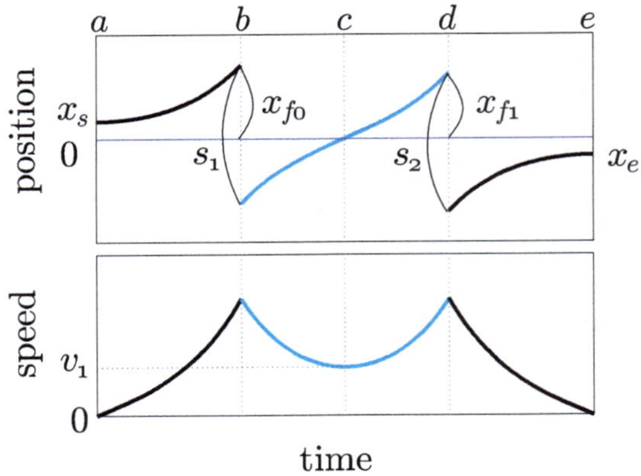

Fig. 4.10 Planned trajectory of the center of mass, position and velocity

Since speed at each exchange are calculated by (4.19), the complete trajectory of each step can be obtained. Figure 4.10 shows the time profile of the position and the velocity of the CoM. The position graph takes origin at each support point, therefore, the plot jumps at the moment of support exchange (b, d) and the amount of the jump indicates the step length. The velocity graph depicts the dynamic change of CoM speed and the peaks at support exchange.

4.2.6 Extension to a Walk on Uneven Terrain

Although the method explained so far was limited to a walk on level ground, we can use the same method for walking on uneven terrain with a small modification. Let us explain this.

Returning to the inverted pendulum model of Fig. 4.2, we consider the case that the center of mass moves on a sloped line as illustrated in Fig. 4.11 described by

$$z = kx + z_c \tag{4.20}$$

where k is the inclination of the line, z_c is the z intersection of the line. We call the line which the CoM moves along as *constraint line.*

Let us calculate the kick force f to realize such motion. First, we decompose the kick force f into the horizontal part f_x and the vertical part f_z

$$f_x = f \sin\theta = (x/r)f \tag{4.21}$$

$$f_z = f \cos\theta = (z/r)f. \tag{4.22}$$

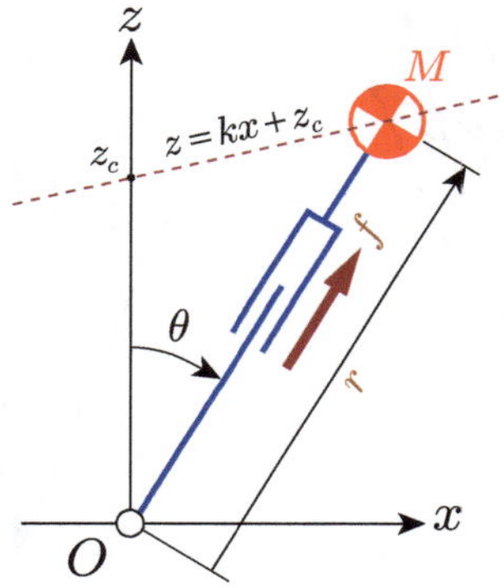

Fig. 4.11 The CoM is controlled to move on the constraint line (broken line) by the kick force f, while the ankle rotates freely($\tau = 0$).

To let the CoM move along the constraint line, the sum of the kick force and the gravity force must be parallel with the constraint line. That is explained as

$$f_x : f_z - Mg = 1 : k. \tag{4.23}$$

Substituting (4.21) and (4.22) into (4.23) and solving for f, we obtain

$$f = \frac{Mgr}{z - kx}. \tag{4.24}$$

A simpler equation can be obtained by using the constraint equation (4.20)

$$f = \frac{Mgr}{z_c}. \tag{4.25}$$

Therefore, the CoM moves on a constraint line by applying the kick force f in proportional with the leg length r (in addition, we need the initial condition matching the constraint). Such pendulum motions are illustrated in Fig. 4.12. In this graph, the arrows are indicating the magnitude and direction of the kick force.

Let us see the dynamics of the CoM under this control. The horizontal dynamics of the CoM can be obtained by substituting (4.25) into (4.21), and applying $f_x = M\ddot{x}$ to obtain

$$\ddot{x} = \frac{g}{z_c}x. \tag{4.26}$$

Surprisingly, this is identical to (4.4), which was derived for the horizontal motion of CoM! To make clear the meaning of this result, we display in Fig. 4.13 the simulation of two inverted pendula moving on different constraint lines. The two constraint lines have different slopes but the same intersection z_c, and the pendula start from the same horizontal position x_0. One might expect that the pendulum on an ascending line ($k > 0$) will decelerate

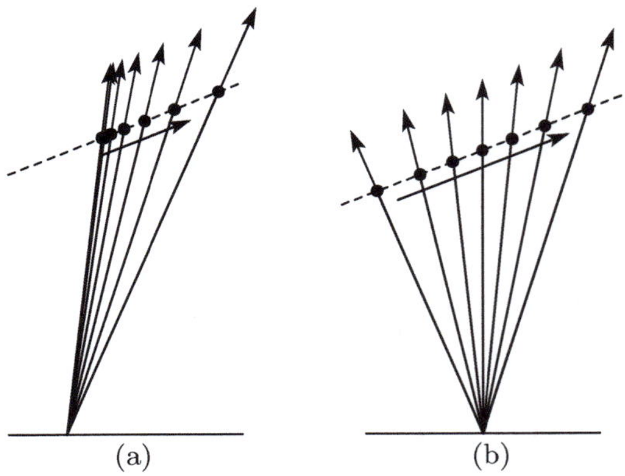

Fig. 4.12 Kick force vector and pendulum motion

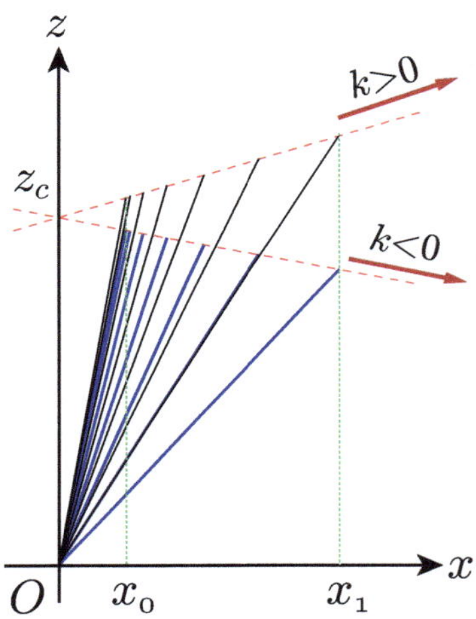

Fig. 4.13 Horizontal motion of a linear inverted pendulum is independent from the slope of the constraint line.

and the pendulum on an descending line ($k < 0$) will accelerate. Nevertheless, the horizontal motions of two pendula are exactly same and they arrive at x_1 simultaneously as shown in Fig. 4.13. This is because the gravity effect is canceled by the kick force and the pendulum motion is governed only by the horizontal location of the CoM.

By using this fact, we can easily design a walking pattern on uneven terrain. Fig. 4.14 shows an example planning of stair climbing. First, we set proper landing points (triangles), then specify constraint lines by connecting the points which are located at the height of z_c from the landing points. By controlling the CoM to move along these constraint lines, we get the horizontal dynamics of CoM for each step as

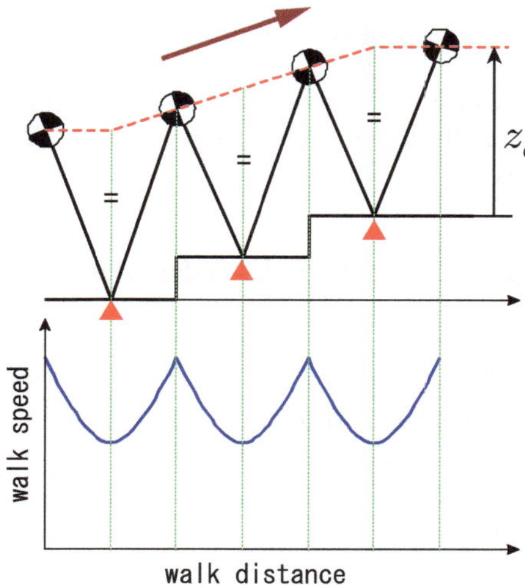

Fig. 4.14 Walking on the stairs by using linear inverted pendulum. (above) Setting of constraint lines (lower) horizontal velocity of the center of mass.

$$\ddot{x} = \frac{g}{z_c}x.$$

Thus, we can apply the method of the previous section. The lower graph of Fig. 4.14 indicates that the horizontal speed of the CoM is not affected by the stair climbing.

By these method, we can easily produce walking patterns for various ground surface geometries. The authors have developed a simple biped robot (Fig. 4.15) which could walk over stairs and obstacles by real-time sensing [116].

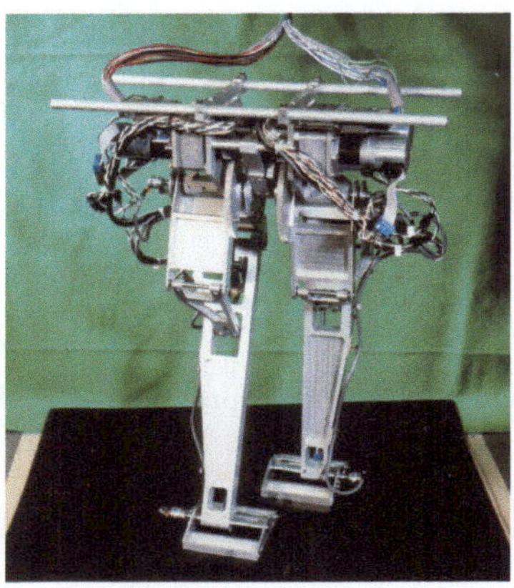

Fig. 4.15 Meltran II, a biped robot with bird-like legs. The light weight legs allows us to treat this robot as a linear inverted pendulum.

4.3 3D Walking Pattern Generation

In this section, we extend the linear inverted pendulum to 3D and examine its nature. Using a 3D linear inverted pendulum, a 3D walking pattern generation is described.

4.3.1 3D Linear Inverted Pendulum

Let us approximate a biped walking robot in 3D space as an imaginary inverted pendulum of Fig. 4.16, which consists of the CoM of the robot and a massless leg connecting the CoM and the supporting point. We assume the pendulum can freely rotate about the supporting point and the leg can change its length by using a kick force f. We can decompose the kick force f into three components, x,y and z

$$f_x = (x/r)f \tag{4.27}$$

$$f_y = (y/r)f \tag{4.28}$$

$$f_z = (z/r)f \tag{4.29}$$

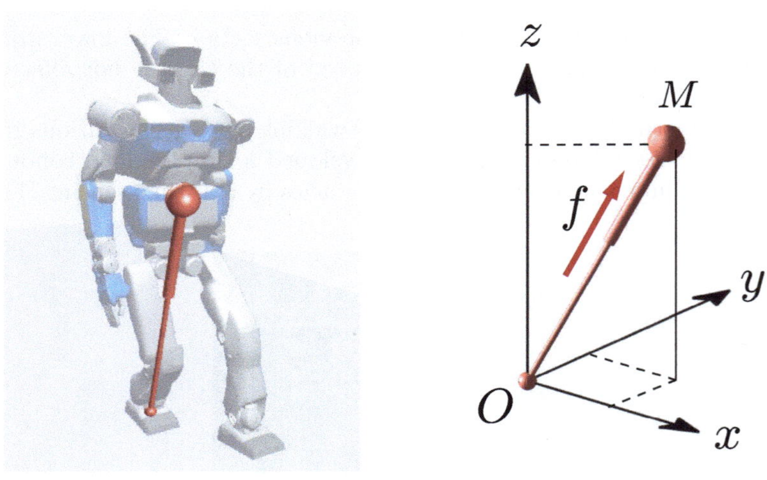

Fig. 4.16 3D inverted pendulum as an approximated walking robot. The supporting point is a spherical joint which allows free rotation. The leg can change its length by generating a kick force f.

where r is the distance between the supporting point and the CoM. Only the kick force and gravity act on the CoM, thus, the motion equation of the CoM is given

$$M\ddot{x} = (x/r)f \tag{4.30}$$
$$M\ddot{y} = (y/r)f \tag{4.31}$$
$$M\ddot{z} = (z/r)f - Mg. \tag{4.32}$$

As we did for 2D inverted pendulum, let us consider a constraint for the CoM. For a 3D inverted pendulum, we introduce a *constraint plane* defined as

$$z = k_x x + k_y y + z_c, \tag{4.33}$$

where k_x, k_y determine the slope and z_c determines the height of the constraint plane.

To let the CoM move along this plane, we need its acceleration be orthogonal to the normal vector of the constraint. Therefore, we need

$$[f(\frac{x}{r}) \quad f(\frac{y}{r}) \quad f(\frac{z}{r}) - Mg] \begin{bmatrix} -k_x \\ -k_y \\ 1 \end{bmatrix} = 0. \tag{4.34}$$

By solving this equation for f, and substituting into (4.33), we obtain

$$f = \frac{Mgr}{z_c}. \tag{4.35}$$

The center of mass moves on the constraint plane by applying the kick force f in proportion to the leg length r. Figure 4.17 shows an image of this motion.

The horizontal dynamics of the CoM can be derived from (4.30) and (4.31) by substituting the kick force of (4.35) to obtain

$$\ddot{x} = \frac{g}{z_c}x, \tag{4.36}$$
$$\ddot{y} = \frac{g}{z_c}y. \tag{4.37}$$

These are linear equations having z_c, the intersection of the constraint plane as the only parameter. The inclination parameters k_x, k_y of the constraint plane do not affect the horizontal motion of the CoM since they are not part of (4.36) and (4.37). We call such pendulum as *3D linear inverted pendulum*[4].

[4] The 3D linear inverted pendulum was originally discovered by Hara, Yokogawa and Sadao [59]. It was expanded to take into account of the ankle torque by Kajita, Matsumoto and Saigo [118].

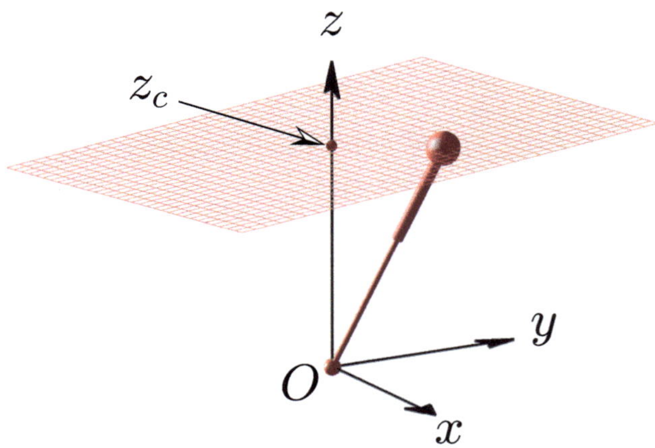

Fig. 4.17 3D linear inverted pendulum. The center of mass moves on the constraint plane by properly controlling the kick force. Inclination of the constraint plane does not affect to the horizontal motion of the CoM.

4.3.2 Natures of the 3D Linear Inverted Pendulum

The nature of 3D linear inverted pendulum is much more interesting although it is a concatenation of two 2D linear inverted pendula. Figure 4.18 shows three trajectories of a 3D linear inverted pendulum whose constraint plane

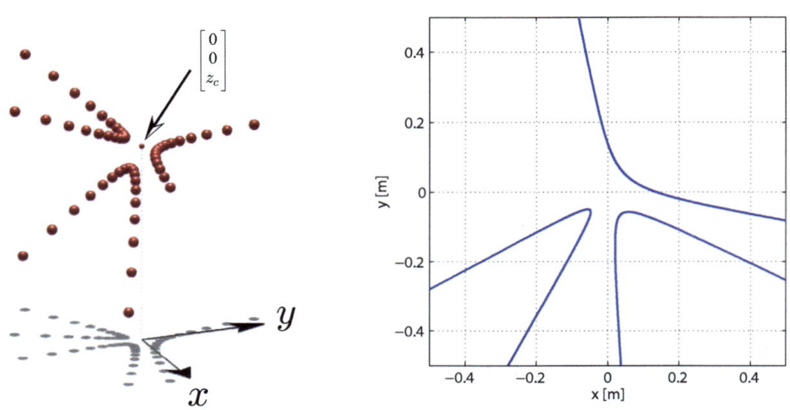

(a) (b)

Fig. 4.18 Motions of 3D linear inverted pendulum. (a)CoM trajectories in 3D space (b)Horizontal projections of trajectories.

has the same z intersection. We can observe an impressive comet-like trajectory having the z intersection $[0\ 0\ z_c]$ as the focal point[5]. We can treat the dynamics of these three-dimensional trajectories as the projection onto the xy-plane (Fig. 4.18(b)) by neglecting the inclination of the constraint plane.

1 Kepler's Second Law

From an analogy from celestial mechanics, let us check Kepler's second law: A planet has constant areal speed. Areal velocity v_{area} is area swept by the line connecting origin and the CoM per unit time and is given as

$$v_{area} = \frac{1}{2}(x\dot{y} - \dot{x}y). \tag{4.38}$$

We can check the time derivative of areal speed for 3D linear inverted pendulum as follows

$$\begin{aligned}
\frac{d}{dt}(v_{area}) &= \frac{1}{2}(\dot{x}\dot{y} + x\ddot{y} - \ddot{x}y - \dot{x}\dot{y}) \\
&= \frac{1}{2}(x\ddot{y} - \ddot{x}y) \quad \leftarrow \text{(substitute eqs.(4.36) and (4.37))} \\
&= \frac{1}{2}(x\frac{g}{z_h}y - \frac{g}{z_h}xy) \\
&= 0.
\end{aligned}$$

Thus, the areal speed of 3D linear inverted pendulum is constant[6].

2 Rotation of the Reference Frame

Next we consider coordinate transformation from the original reference frame xy to a new reference frame $x'y'$ rotated θ from it as in Fig. 4.19. The transformation is

$$x = cx' - sy' \tag{4.39}$$

$$y = sx' + cy' \tag{4.40}$$

$$c \equiv \cos\theta, s \equiv \sin\theta.$$

[5] LIP trajectories are different from comet trajectories which travel around the Sun. More proper analogy is a trajectory of particles in Rutherford scattering experiment which proved the existence of the nucleus of an atom. As shown subsequently, this is indeed a very good analogy.

[6] Generally, areal speed is conserved by any motion in a central force field. It represents the conservation of angular momentum in a different way.

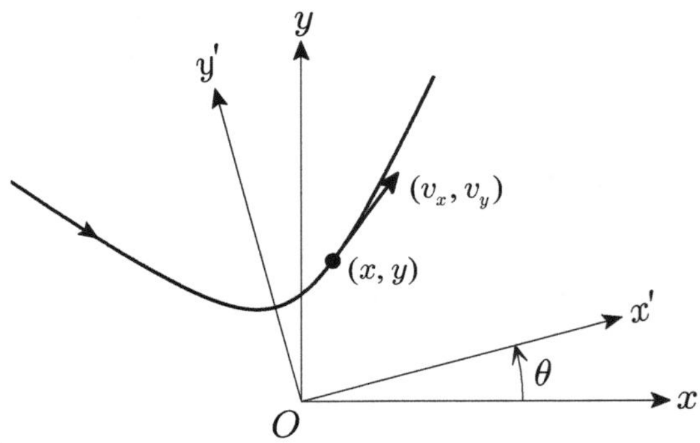

Fig. 4.19 Horizontal projection of a 3D-LIPM trajectory and the new reference frame $x'y'$ rotated θ from the original frame

Let us confirm that the equations of the 3D linear inverted pendulum (4.36) and (4.37) are still valid in the new frame $x'y'$. By substituting the coordinate transformation given by above equations we get

$$c\ddot{x}' - s\ddot{y}' = \frac{g}{z_h}(cx' - sy') \tag{4.41}$$

$$s\ddot{x}' + c\ddot{y}' = \frac{g}{z_h}(sx' + cy'). \tag{4.42}$$

From $c\times$(4.41) $+s\times$(4.42) and $-s\times$(4.41) $+c\times$(4.42) we get the dynamics represented in $x'y'$ frame given by

$$\ddot{x}' = \frac{g}{z_c}x'$$

$$\ddot{y}' = \frac{g}{z_c}y'.$$

This transformation is possible for any given rotation angle θ. Therefore, we have confirmed that the 3D linear inverted pendulum dynamics can be always separated into two orthogonal components independently from the orientation of the reference frame[7].

3 Geometry of 3D Linear Inverted Pendulum Trajectory

We can calculate the geometry of the trajectory using the transformation of Fig. 4.19 and orbital energies. The orbital energy along the x' axis of a new frame is given as

[7] Motions under gravity force or electro-static force, which is proportional to inverse square of the distance cannot be treated like this.

$$E'_x = -\frac{g}{2z_c}(cx + sy)^2 + \frac{1}{2}(c\dot{x} + s\dot{y})^2. \quad (4.43)$$

This orbital energy E'_x changes by the rotation of the frame θ and takes extreme value when x' or y' correspond to the axis of symmetry of the trajectory. From

$$\begin{aligned}\frac{\partial E'_x}{\partial \theta} &= -A\cos 2\theta + \frac{B}{2}\sin 2\theta = 0 \\ A &\equiv (g/z_c)xy - \dot{x}\dot{y} \\ B &\equiv (g/z_c)(x^2 - y^2) - (\dot{x}^2 - \dot{y}^2)\end{aligned} \quad (4.44)$$

we can calculate θ which indicates the symmetry axis of the trajectory,

$$\theta = \begin{cases} (1/2)\tan^{-1}(2A/B) & (\text{if } B \neq 0) \\ \pi/4 & (\text{if } A \neq 0, B = 0). \end{cases} \quad (4.45)$$

In the case of $A = 0, B = 0$, the trajectory becomes a straight line toward origin or a straight line from origin, and the line itself is a symmetry axis.

Suppose we have already chosen the frame x, y to fit the axis of symmetry. In this case, the orbital energy E'_x must take extreme value with $\theta = 0$. By substituting $\theta = 0$ into (4.44), we get

$$(g/z_c)xy - \dot{x}\dot{y} = 0.$$

After transferring the second term to the right side, we square both sides of the equation to obtain

$$(g/z_c)^2 x^2 y^2 = \dot{x}^2 \dot{y}^2.$$

We substitute the following two equations which were derived from the definition of orbital energy

$$\dot{x}^2 = 2E_x + \frac{g}{z_c}x^2 \quad \dot{y}^2 = 2E_y + \frac{g}{z_c}y^2$$

to obtain an equation which represents geometric form of 3D linear inverted pendulum trajectory given by

$$\frac{g}{2z_c E_x}x^2 + \frac{g}{2z_c E_y}y^2 + 1 = 0. \quad (4.46)$$

This is an equation of hyperbola since one of E_x, E_y is negative and the other is positive[8].

[8] Hyperbolic trajectories are observed in Rutherford scattering or swing-by trajectories of spacecraft like Voyager I and II. I feel excitement finding similar trajectories in elementary particles, in planetary space, and in biped locomotion. Don't you?!

4.3.3 3D Walking Pattern Generation

Figure 4.20 shows an example of walking pattern based on the 3D linear inverted pendulum. By setting a proper constraint plane, we can apply the same pattern to stairs or an uneven floor.

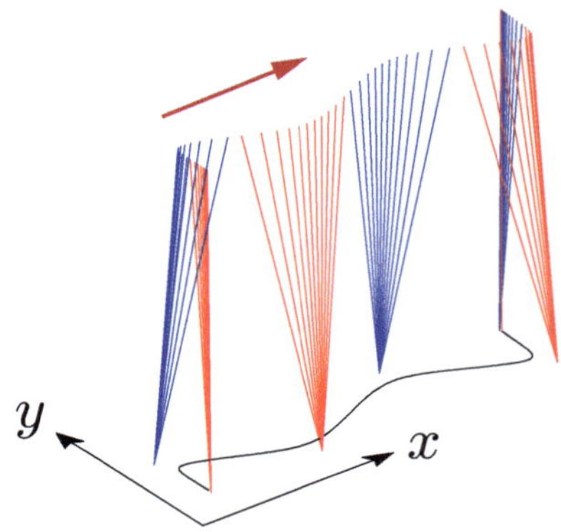

Fig. 4.20 Walking pattern on flat floor based on 3D linear inverted pendulum. Three forward steps from a standstill to a stop.

For three dimensional walking, we need a simultaneous support exchange for the x and y directions, thus we cannot use the walking pattern generation explained in Section 4.2.5, which requires an arbitrary time of support exchange. In the following discussion, we consider walking with constant pace of support exchange and denote T_{sup} for support period of each step.

1 Walk Primitive

Let us define a piece of a 3D linear inverted pendulum trajectory as Fig. 4.21, which is symmetric about y axis and defined in a period of $[0 \quad T_{sup}]$. As previously described, Fig. 4.21(a) is a part of hyperbola. We will call this piece of trajectory as a *walk primitive*.

When a support time T_{sup} and an intersection of constraint plane z_c are given, a walk primitive is uniquely determined by its terminal position $(\bar{x}, \bar{y})$ since it is symmetric. The terminal speed $(\bar{v}_x, \bar{v}_y)$ can be calculated as follows.

Using the symmetric nature of the walk primitive, the initial condition along the x axis is $(-\bar{x},\ \bar{v}_x)$ and the terminal position is $\bar{x}$. From the analytic solution of linear inverted pendulum (4.5), we have

$$\bar{x} = -\bar{x}C + T_c\bar{v}_xS \tag{4.47}$$

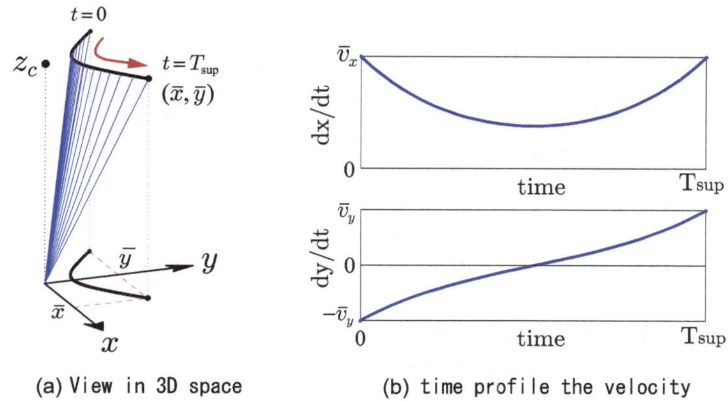

Fig. 4.21 Walk primitive: basic 3D walking pattern

where

$$T_c \equiv \sqrt{\frac{z_c}{g}},\ C \equiv \cosh\frac{T_{sup}}{T_c},\ S \equiv \sinh\frac{T_{sup}}{T_c}.$$

Solving (4.47) for the terminal velocity $\bar{v}_x$ gives

$$\bar{v}_x = \bar{x}(C+1)/(T_c S). \tag{4.48}$$

Likewise, for the y component of the walk primitive, the initial condition is $(\bar{y},\ -\bar{v}_y)$ and the terminal position is $\bar{y}$, thus we get the terminal velocity as,

$$\begin{aligned} \bar{y} &= \bar{y}C + T_c(-\bar{v}_y)S, \\ \bar{v}_y &= \bar{y}(C-1)/(T_c S). \end{aligned} \tag{4.49}$$

Walking primitives allows us to easily produce a walking trajectory. For example, a straight walk with step length of $2\bar{x}$ can be made by connecting identical walk primitives, reversing the sign of the y-component after each contact.

2 Walk Parameters

In practical situations like stair climbing or obstacle avoidance, it is frequently required to directly specify the foot placements. Figure 4.22 illustrates an example of simple foot placements, which can be represented by using step length and step width as the following data.

n	1	2	3	4	5
$s_x^{(n)}$	0.0	0.3	0.3	0.3	0
$s_y^{(n)}$	0.2	0.2	0.2	0.2	0.2

where s_x are the step length along the walking direction, and s_y are the step width for lateral direction. We call this data the *walk parameters* since it is

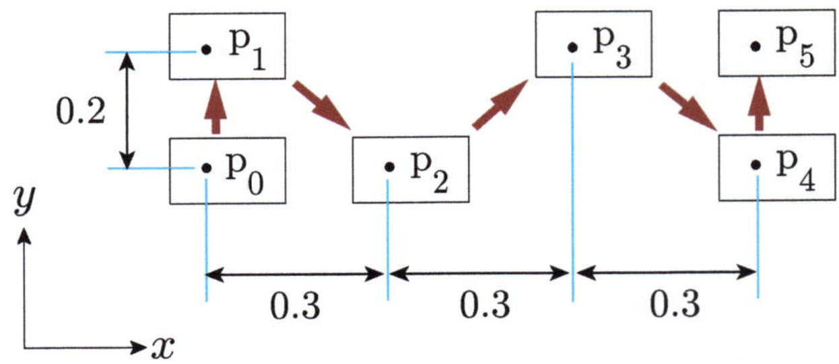

Fig. 4.22 Points of foot placement $p_0 \ldots p_N$. Step length and width are determined from these points. The rectangles are footprints of a robot.

encoding the foot placements in Fig. 4.22. By putting superscript (n) to indicate the data of n-th step, the n-th footplace $(p_x^{(n)}, p_y^{(n)})$ can be represented as

$$\begin{bmatrix} p_x^{(n)} \\ p_y^{(n)} \end{bmatrix} = \begin{bmatrix} p_x^{(n-1)} + s_x^{(n)} \\ p_y^{(n-1)} - (-1)^n s_y^{(n)} \end{bmatrix}, \tag{4.50}$$

where $(p_x^{(0)}, p_y^{(0)})$ is the place of the first support foot, in this case, the right foot. If the first support foot was the left foot, we must replace $-(-1)^n$ by $+(-1)^n$ in the equation.

The walk primitive for the n-th step can be determined as

$$\begin{bmatrix} \bar{x}^{(n)} \\ \bar{y}^{(n)} \end{bmatrix} = \begin{bmatrix} s_x^{(n+1)}/2 \\ (-1)^n s_y^{(n+1)}/2 \end{bmatrix}. \tag{4.51}$$

Note that the walk primitive of the n-th step is determined by the $n+1$-th step length and width. This is necessary for the proper coordination between foot placements and walking motion.

The terminal velocity of a walk primitive is calculated by (4.48) and (4.49),

$$\begin{bmatrix} \bar{v}_x^{(n)} \\ \bar{v}_y^{(n)} \end{bmatrix} = \begin{bmatrix} (C+1)/(T_c S)\bar{x}^{(n)} \\ (C-1)/(T_c S)\bar{y}^{(n)} \end{bmatrix}. \tag{4.52}$$

The series of walk primitives determined in this way are discontinuous at the beginning and the end of a set of steps. The method to obtain a continuous and realizable walking pattern is explained in the next section.

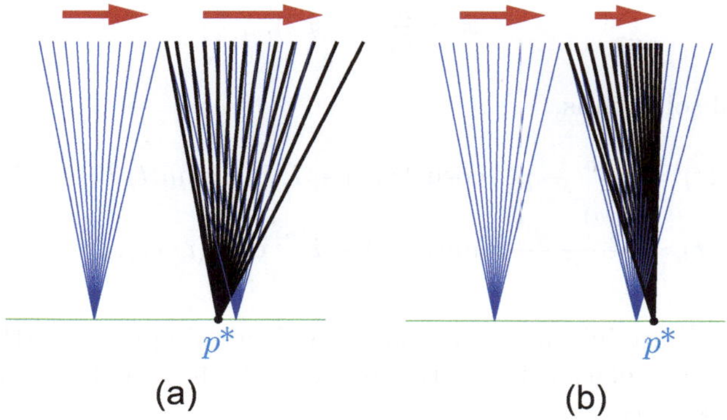

Fig. 4.23 Walking speed adjustment for fixed step cycle [41, 97] (a) Speed up by taking a shorter step. (b) Slow down by taking a longer step.

3 Modification of Foot Placements

For a robot walking with fixed step cycle, we can control its speed by adjusting foot placements[9]. Its intuitive explanation is depicted in Fig. 4.23.

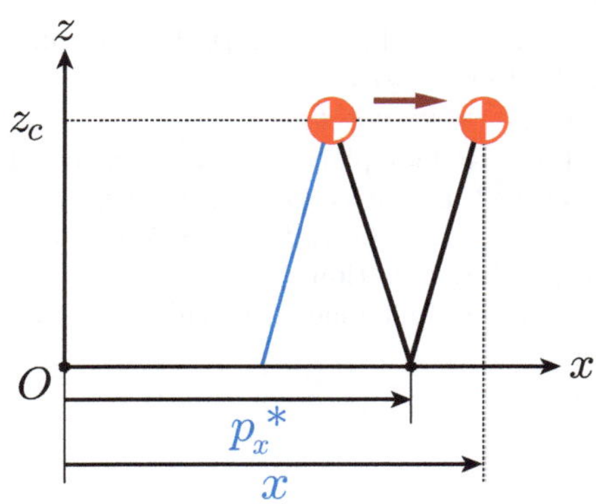

Fig. 4.24 A Linear Inverted Pendulum represented in the ground fixed frame

Let us denote the modified foot placement as p_x^* and calculate how it affects the walking motion. Figure 4.24 illustrates a linear inverted pendulum with respect to the ground fixed frame[10],

[9] In the control method of Fig. 4.7 (page.114), the step length was fixed, and the walking speed was controlled by modifying touch down timing.

[10] We will only discuss the motion along x-axis, but the same result is obtained for y-axis motion.

$$\ddot{x} = \frac{g}{z_c}(x - p_x^*). \tag{4.53}$$

Its analytical solution is,

$$x(t) = (x_i^{(n)} - p_x^*)\cosh(t/T_c) + T_c\dot{x}_i^{(n)}\sinh(t/T_c) + p_x^* \tag{4.54}$$

$$\dot{x}(t) = \frac{x_i^{(n)} - p_x^*}{T_c}\sinh(t/T_c) + \dot{x}_i^{(n)}\cosh(t/T_c), \tag{4.55}$$

where $x_i^{(n)}, \dot{x}_i^{(n)}$ are initial conditions at the beginning of the n-th step.

Therefore, the relationship between the foot placement p_x^* and the final state of n-th step is

$$\begin{bmatrix} x_f^{(n)} \\ \dot{x}_f^{(n)} \end{bmatrix} = \begin{bmatrix} C & T_c S \\ S/T_c & C \end{bmatrix} \begin{bmatrix} x_i^{(n)} \\ \dot{x}_i^{(n)} \end{bmatrix} + \begin{bmatrix} 1 - C \\ -S/T_c \end{bmatrix} p_x^*. \tag{4.56}$$

Step 1 Set support period T_{sup} and walk parameters s_x, s_y. Set initial position of CoM (x, y) and initial foot placement $(p_x^*, p_y^*) = (p_x^{(0)}, p_y^{(0)})$.
Step 2 $T := 0, n := 0$.
Step 3 Integrate equation of linear inverted pendulum (4.53) (and the equation for y-axis) from T to $T + T_{sup}$.
Step 4 $T := T + T_{sup}, \quad n := n + 1$
Step 5 Calculate the next foot place $(p_x^{(n)}, p_y^{(n)})$ using (4.50).
Step 6 Set the next walk primitive $(\bar{x}^{(n)}, \bar{y}^{(n)})$ using (4.51) and (4.52).
Step 7 Calculate target state $(x^d, \dot{x}^d)$ by (4.57). Calculate target state $(y^d, \dot{y}^d)$ by corresponding equation.
Step 8 Calculate modified foot placement (p_x^*, p_y^*) by (4.59) (as well as y component).
Step 9 Goto Step 3.

Fig. 4.25 Algorithm of walking pattern generation based on 3D-LIP

As the target state we use the terminal state of the walk primitive presented in the ground frame

$$\begin{bmatrix} x^d \\ \dot{x}^d \end{bmatrix} = \begin{bmatrix} p_x^{(n)} + \bar{x}^{(n)} \\ \bar{v}_x^{(n)} \end{bmatrix}. \tag{4.57}$$

Let us calculate the foot placement which ends up the final state closest to the target $(x^d, \dot{x}^d)$. The evaluation function can be defined as

$$N \equiv a(x^d - x_f^{(n)})^2 + b(\dot{x}^d - \dot{x}_f^{(n)})^2 \tag{4.58}$$

where a, b are positive weights. Substituting (4.56) into the evaluation function and by using $\partial N/\partial p_x^* = 0$, we can obtain the foot placement which will minimize N

$$\begin{aligned} p_x^* &= -\frac{a(C-1)}{D}(x^d - Cx_i^{(n)} - T_c S\dot{x}_i^{(n)}) \\ &\quad -\frac{bS}{T_c D}(\dot{x}^d - \frac{S}{T_c}x_i^{(n)} - C\dot{x}_i^{(n)}) \\ D &\equiv a(C-1)^2 + b(S/T_c)^2. \end{aligned} \tag{4.59}$$

The method of walking pattern generation can be organized as an algorithm shown as Fig. 4.25.

Figure 4.26 shows the generated walking pattern based on the walk parameters of Fig. 4.22. We can observe a slight back step at the walk start to obtain acceleration, and a slightly wider step at the walk end to obtain deceleration. Since such modifications of foot placement are necessary, this method cannot realize the exact foot placement specified by the walk parameters. Nevertheless, since (4.59) guarantees the error converges to zero, the robot can take a specified foot placement in steady walking.

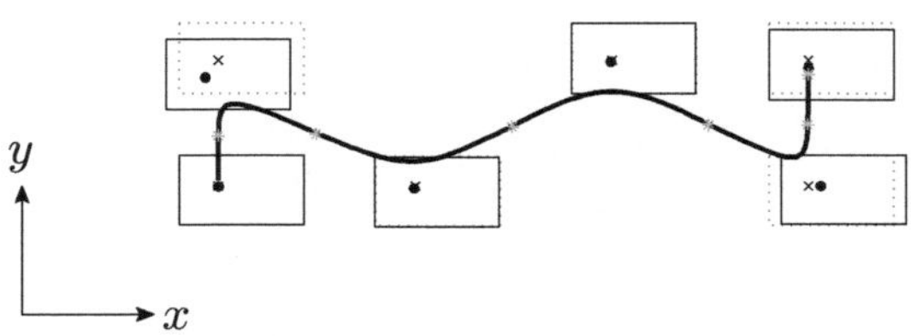

Fig. 4.26 Walking pattern generated by the proposed algorithm. Bold line is the trajectory of the CoM, black circles are modified foot placements. The desired foot placements are shown by $\times$. For walk start and walk end, foot placements are modified largely to obtain proper acceleration and deceleration. $z_c = 0.8, T_{sup} = 0.8$,$a = 10, b = 1$.

For diagonal walking, we change s_y for each step.

n	1	2	3	4	5
$s_x^{(n)}$	0.0	0.2	0.2	0.2	0
$s_y^{(n)}$	0.2	0.3	0.1	0.3	0.2

Figure 4.27 shows the pattern generated by this set of walk parameters. If we set all s_x to zero in this set, pure side walking is realized.

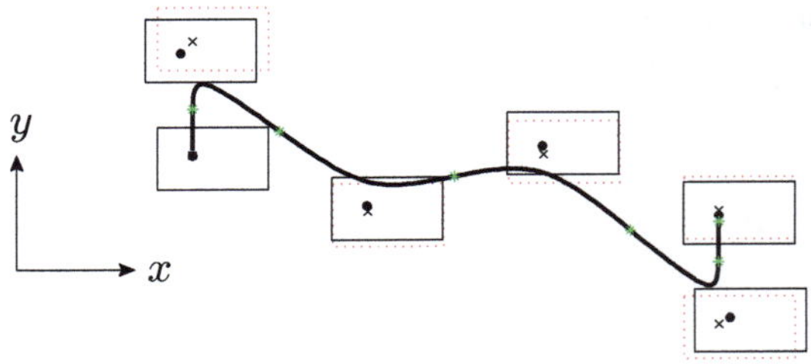

Fig. 4.27 Example of diagonal walk. To obtain a side step, s_y is changed for each step. $z_c = 0.8, T_{sup} = 0.8,\ a = 10, b = 1.$

4 Changing Walk Direction

For changing walk direction, we need to add heading information to our walk parameters. The direction change taking a step is denoted by s_θ as illustrated in Fig. 4.28[11].

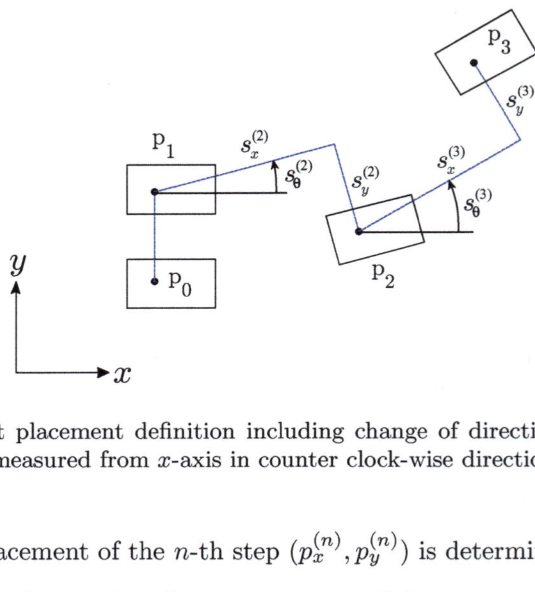

Fig. 4.28 Foot placement definition including change of direction. The heading direction s_θ is measured from x-axis in counter clock-wise direction.

The foot placement of the n-th step $(p_x^{(n)}, p_y^{(n)})$ is determined by

$$\begin{bmatrix} p_x^{(n)} \\ p_y^{(n)} \end{bmatrix} = \begin{bmatrix} p_x^{(n-1)} \\ p_y^{(n-1)} \end{bmatrix} + \begin{bmatrix} \cos s_\theta^{(n)} & -\sin s_\theta^{(n)} \\ \sin s_\theta^{(n)} & \cos s_\theta^{(n)} \end{bmatrix} \begin{bmatrix} s_x^{(n)} \\ -(-1)^n s_y^{(n)} \end{bmatrix}. \tag{4.60}$$

[11] This is a little bit simpler definition than the method we are using now. For example, this method cannot generate a stepping turn in place. Finding a better definition is an exercise for the readers.

The walk primitive for the n-th step is given by

$$\begin{bmatrix} \bar{x}^{(n)} \\ \bar{y}^{(n)} \end{bmatrix} = \begin{bmatrix} \cos s_\theta^{(n+1)} & -\sin s_\theta^{(n+1)} \\ \sin s_\theta^{(n+1)} & \cos s_\theta^{(n+1)} \end{bmatrix} \begin{bmatrix} s_x^{(n+1)}/2 \\ (-1)^n s_y^{(n+1)}/2 \end{bmatrix}. \tag{4.61}$$

The speed of walk primitive is also calculated as

$$\begin{bmatrix} \bar{v}_x^{(n)} \\ \bar{v}_y^{(n)} \end{bmatrix} = \begin{bmatrix} \cos s_\theta^{(n+1)} & -\sin s_\theta^{(n+1)} \\ \sin s_\theta^{(n+1)} & \cos s_\theta^{(n+1)} \end{bmatrix} \begin{bmatrix} (1+C)/(T_c S)\bar{x}^{(n)} \\ (C-1)/(T_c S)\bar{y}^{(n)} \end{bmatrix}. \tag{4.62}$$

One can generate a walking pattern with arbitrary heading control by using (4.60) instead of (4.50) in walking pattern generation algorithm Step 5, and by using (4.61) and (4.62) instead of (4.51) and (4.51) in the Step 6.

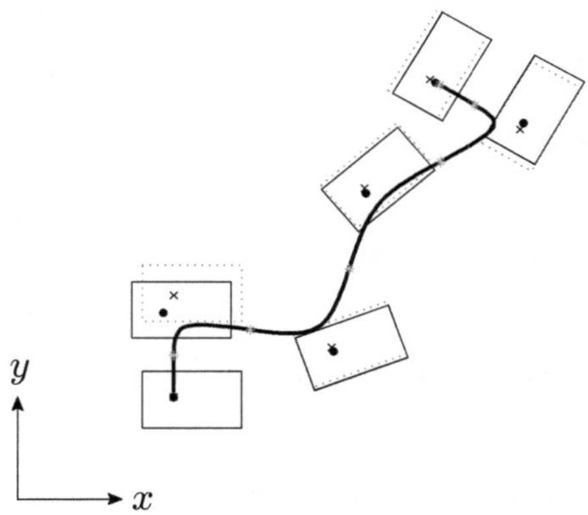

Fig. 4.29 Walk on an arc. 20 [deg] rotation per step. $z_c = 0.8, T_{sup} = 0.8$,weights $a = 10, b = 1$.

For example, to create a walk on an arc we can use the following parameters which specifies 20 [deg] rotation per one step.

n	1	2	3	4	5
s_x	0.0	0.25	0.25	0.25	0
s_y	0.2	0.2	0.2	0.2	0.2
s_θ	0	20	40	60	60

Figure 4.29 shows the walking trajectory generated from this walk parameter.

4.3.4 Introducing Double Support Phase

So far, we have been assuming that the support leg of the inverted pendulum model is exchanged instantaneously. However, such abrupt support

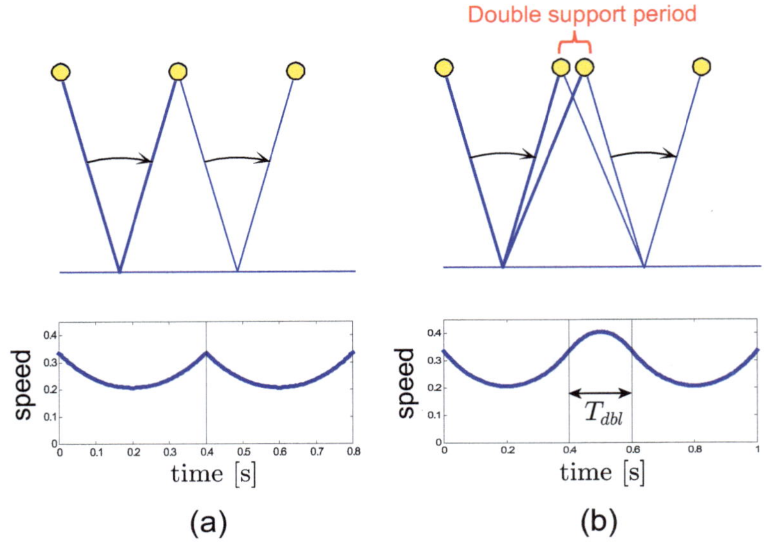

Fig. 4.30 Improvement of support exchange. (a) Instantaneous support exchange results bending point of CoM speed graph. CoM acceleration jumps from maximum to minimum. (b) By introducing double support phase, we can obtain continuous profiles of CoM speed and acceleration.

exchange results the horizontal acceleration jumps from maximum to minimum Fig. 4.30(a). As the result, the robot may suffer huge impacts and may possibly be damaged.

To obtain a smoother walking pattern which is suitable for real robots, a double support phase with a predetermined period T_{dbl} is inserted in the moment of support leg exchange.

What we need is smooth velocity profiles without sudden changes of slope. By such motion, we can avoid discontinuous changes of acceleration, thus the ZMP smoothly transfers from the former support foot to the new support. For this purpose, we generate velocity profiles using third order polynomials so that we can specify the speeds and accelerations at the beginning and the end of a double support period. As the result, the position of CoM is described by forth order polynomials (Fig. 4.30(b)),

$$x(t) = a_0 + a_1 t + a_2 t^2 + a_3 t^3 + a_4 t^4. \tag{4.63}$$

The coefficients $a_0 \ldots a_4$ are determined by position, velocity and acceleration of the CoM at the instant of support exchange.

By inserting double supports, a robot takes larger step than planned. This can be compensated by a proportional reduction of walk primitives in advance. Figure 4.31 shows an example of a walking trajectory with a double support phase.

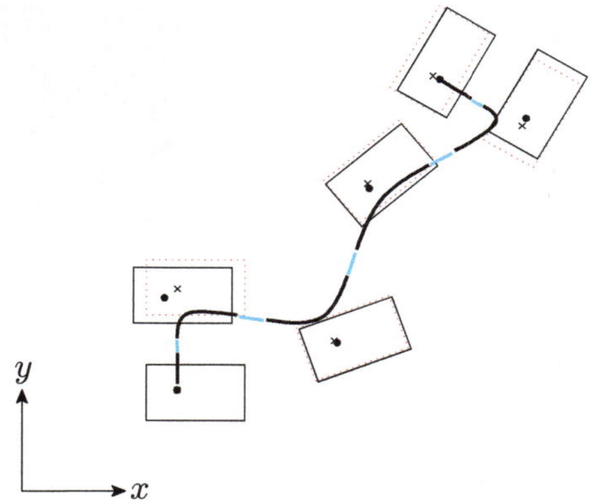

Fig. 4.31 A walking trajectory with double support periods. The CoM trajectories while double support phase are indicated by gray lines. The same parameters as were used in Fig. 4.29 are used. $z_c = 0.8, T_{sup} = 0.7, T_{dbl} = 0.1$,weights $a = 10, b = 1$.

It should be noted that while longer period of double support results smoother support exchange, it also requires undesirable quicker swing leg motion. Therefore we have a trade-off in determining T_{dbl}.

4.3.5 From Linear Inverted Pendulum to Multi-body Model

The easiest way to generate a walking pattern by using the linear inverted pendulum is to let the pelvis link follow the CoM motion of LIP. First, the real position of the CoM is calculated using a multi-body model and its position with respect to the pelvis frame is determined. After that, the position of the pelvis link is directly determined from the liner inverted pendulum assuming that the relative position of the CoM is kept constant with respect to the pelvis. In addition, we must calculate the swing foot trajectory so that it arrives the desired foot place at the specified time of touchdown.

Once we determine the trajectories for the pelvis and the both feet, the leg joint angles can be obtained by inverse kinematics as explained in Chapter 2.

This method is based on an assumption that the multi-body dynamics of the robot can be approximated by a simple inverted pendulum and its validity can be confirmed by using ZMP described in the former chapter. By calculating ZMP using multi-body model, we can evaluate the effects of swing leg reaction and errors in CoM position which were neglected in a linear inverted pendulum. Figure 4.32 shows two ZMPs, one based on the linear inverted pendulum, and one based on multi-body dynamics and the proposed

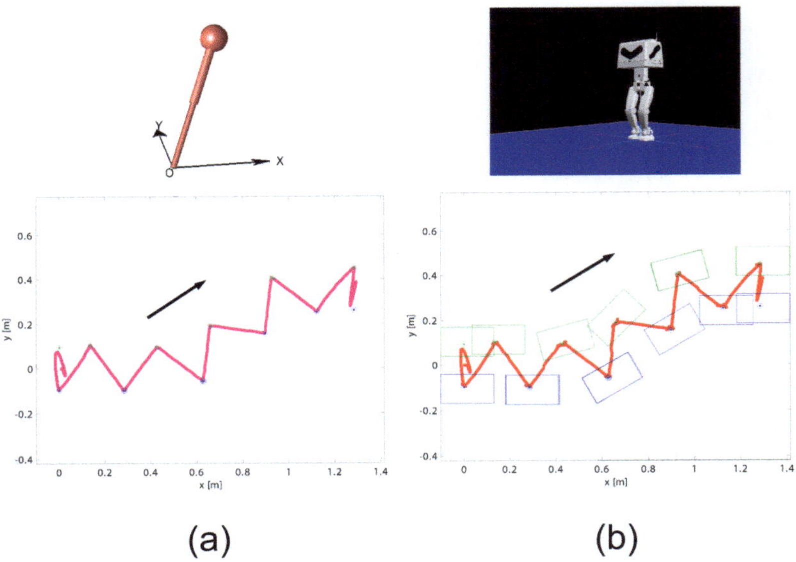

Fig. 4.32 Comparison of ZMP trajectory (a) ZMP calculated from 3D linear inverted pendulum model (b) ZMP calculated from multi-body dynamics whose pelvis link moves as 3D-LIP

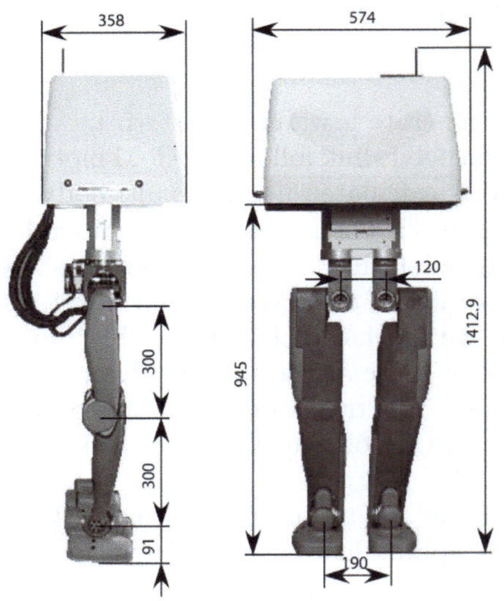

Fig. 4.33 Biped robot HRP-2L

pattern generation. Both of ZMPs are sufficiently close, hence we can conclude a multi-body dynamics can be simplified as a simple inverted pendulum in this case.

4.3.6 Implementation Example

Let us see an implementation of the proposed walking pattern generation. Figure 4.33 shows a biped robot HRP-2L which was developed in "Humanoid Robotics Project"(HRP). This robot was built to evaluate the leg part of HRP-2, the humanoid robot which was the final goal of the project. Each leg has six degrees of freedom and the robot is equipped with a Pentium II

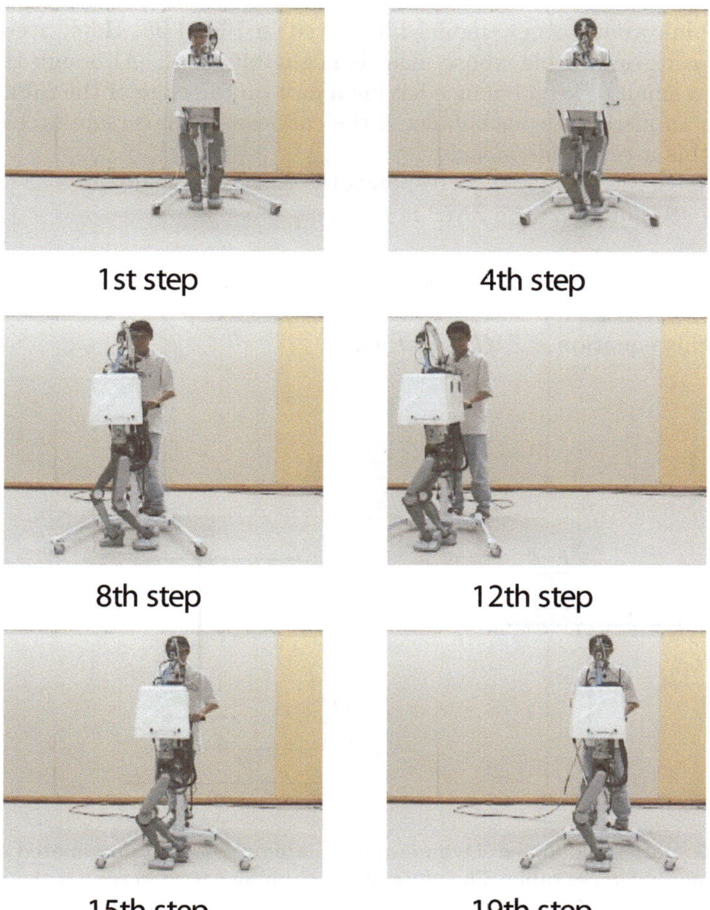

Fig. 4.34 Snapshots of real-time walking control

933MHz based on-board computer on its body part. The total weight is 58.2 [kg] including batteries of 11.4 [kg] and dummy weights of 22.6 [kg] which emulates upper body.

The algorithm of Fig. 4.25 can generate a walking pattern where at least two future steps were given. So we could build a walking control system which allows real-time step modification by specifying the walk parameter of two steps in future (s_x, s_y, s_θ) with a joystick. Figure 4.34 shows snapshots of our experiment of real-time walking control.

4.4 ZMP Based Walking Pattern Generation

4.4.1 Cart-Table Model

Let us think about a new model illustrated in Fig. 4.35. Here, a cart with mass M runs on a table whose mass is negligibly small. Although the table foot is too small to keep balance having a cart on the edge of the table, it can still keep an instantaneous balance if the cart runs with certain acceleration. We call this a *cart-table model.*

Since a cart-table model corresponds the case of a single mass at constant height in section 3.5.2, the ZMP is given as

$$p = x - \frac{z_c}{g}\ddot{x}. \tag{4.64}$$

We call this equation a *ZMP equation.*

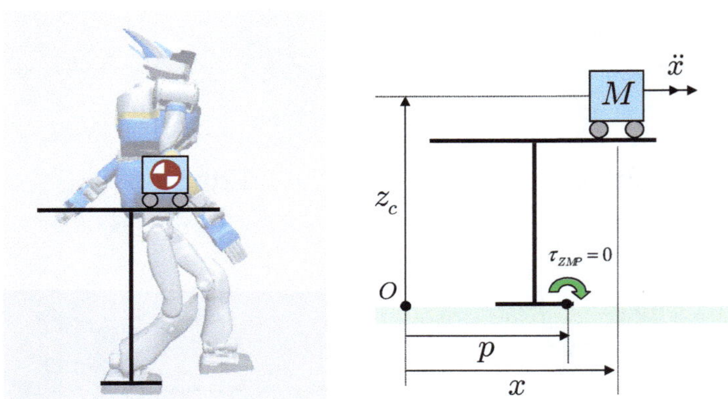

Fig. 4.35 Cart-table model: Dynamics of walking robot is approximated as a cart running on a massless table. The state of the running cart determines the center of pressure which acts from the floor, in other words, the cart changes ZMP.

On the other hand, the equation of linear inverted pendulum was given as following (Fig. 4.24).

$$\ddot{x} = \frac{g}{z_c}(x - p). \tag{4.65}$$

By regarding p as ZMP and not a foot place point as we did previously, we can treat a robot applying ankle torque and a robot in double support phase in a unified manner [134]. Moreover, we can see that (4.64) and (4.65) are the same equations with different outlooks.

A linear inverted pendulum model and a cart-table model are compared in Fig. 4.36. In a linear inverted pendulum model, the CoM motion is generated by the ZMP (Fig. 4.36(a)), and in a cart-table model, the ZMP is generated by the CoM motion (Fig. 4.36(b)). Therefore, these two models have opposite input-output causality.

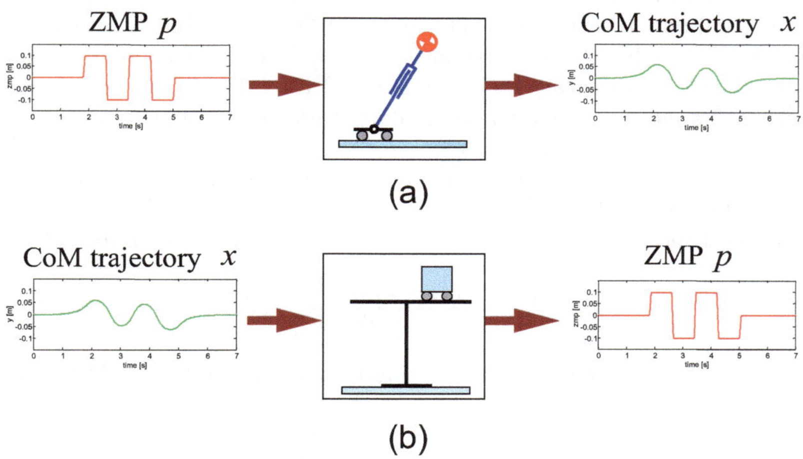

Fig. 4.36 Comparison of two models for relationship between ZMP and CoM (a) A linear inverted pendulum model inputs ZMP and outputs CoM motion. (b) A cart-table model inputs CoM motion and outputs ZMP.

As we described in the former section, a method based on a linear inverted pendulum assumes input-output relationship of Fig. 4.36(a) and the walking pattern is calculated in the following process.

(Specify target CoM motion) $\Rightarrow$ (Calculate appropriate ZMP)

In this case, it is difficult to plan ZMP as expected. Indeed, we have modified the ZMP (support foot placement) in the method of the previous section.

Now, let us consider a walking pattern generation based on a cart-table model. In this case, assuming the causal relationship of Fig. 4.36(b), we calculate a walking pattern by the following manner[12].

(Specify target ZMP trajectory) ⇒ (Calculate appropriate CoM motion)

As the result, we can obtain a walking pattern which realizes the specified ZMP trajectory. Let us call such a method *ZMP based walking pattern generation.*

4.4.2 Off-Line Walking Pattern Generation

ZMP based walking pattern generation was first proposed by Vukobratović and Stepanenko in their paper published in 1972 [90], but their algorithm takes considerable computation time. Then, Takanishi et al. proposed a practical method which transforms the target ZMP pattern into a Fourier series by using FFT, solve the ZMP equation (4.64) in the frequency domain and obtains the CoM trajectory by using inverse FFT [11][13]. A pattern generator based on this method played a particularly important role in the early stage of the Humanoid Robotics Project.

In this section, we introduce a fast and efficient algorithm that was recently proposed by Nishiwaki et al. [114][14] Let us discretize the ZMP equation with a sampling time Δt. For this purpose, the acceleration $\ddot{x}$ is approximated as

$$\ddot{x}_i = \frac{x_{i-1} - 2x_i + x_{i+1}}{\Delta t^2}, \tag{4.66}$$

where $x_i \equiv x(i\Delta t)$. Using this approximation, the discretized ZMP equation is

$$\begin{aligned} p_i &= ax_{i-1} + bx_i + cx_{i+1}, \\ a_i &\equiv -z_c/(g\Delta t^2), \\ b_i &\equiv 2z_c/(g\Delta t^2) + 1, \\ c_i &\equiv -z_c/(g\Delta t^2). \end{aligned} \tag{4.67}$$

Putting the equations (4.67) in a column for the period of the specified $(1 \ldots N)$, and representing them as a single matrix equation gives

[12] There exist an infinite numbers of possible CoM motions which realize the given ZMP trajectory, however, almost all of them suffer divergence. Fig. 4.36(b) can be regarded as a mechanism which guarantees an executable solution.

[13] Later, Takanishi's method was extended to handle real-time pattern generation [40].

[14] Another fast and efficient method was proposed by Nagasaka [93].

$$\begin{bmatrix} p'_1 \\ p_2 \\ \vdots \\ p_{N-1} \\ p'_N \end{bmatrix} = \begin{bmatrix} a_1+b_1 & c_1 & 0 & & & \\ a_2 & b_2 & c_2 & \ddots & & \\ & & \ddots & & & \\ & & \ddots & a_{N-1} & b_{N-1} & c_{N-1} \\ & & & 0 & a_N & b_N+c_N \end{bmatrix} \begin{bmatrix} x_1 \\ x_2 \\ \vdots \\ x_{N-1} \\ x_N \end{bmatrix}, \tag{4.68}$$

where p'_1, p'_N are specified by using initial and terminal velocities v_1, v_N as

$$p'_1 = p_1 + a_1 v_1 \Delta t, \quad p'_N = p_N - c_N v_N \Delta t.$$

Rewriting (4.68) as

$$\boldsymbol{p} = \boldsymbol{A}\boldsymbol{x}$$

gives the representation of the solution by

$$\boldsymbol{x} = \boldsymbol{A}^{-1}\boldsymbol{p}. \tag{4.69}$$

Although $\boldsymbol{A}$ is a huge square matrix with several thousands columns and several thousands lows, there exists an efficient algorithm to compute the inverse [136] since it is a tridiagonal matrix whose elements are all zeros except its main diagonal, the adjacent diagonals above and below it.

From this CoM trajectory, we can generate a walking pattern for a multi-body model by using the method of Section 4.3.5. Then we can calculate the ZMP trajectory for the multi-body model.

$$\boldsymbol{p}^* = RealZMP(\boldsymbol{x}). \tag{4.70}$$

The function $RealZMP()$ calculates ZMP based on a multi-body model and $\boldsymbol{p}^*$ is the obtained ZMP. The ZMP error $\boldsymbol{p}^* - \boldsymbol{p}^d$ contains information about the difference between the cart-table model and the multi-body model. Again using (4.69), we can calculate the CoM variation to compensate the ZMP error

$$\Delta\boldsymbol{x} = \boldsymbol{A}^{-1}(\boldsymbol{p}^* - \boldsymbol{p}^d).$$

The CoM trajectory is updated by

$$\boldsymbol{x} := \boldsymbol{x} - \Delta\boldsymbol{x}.$$

Going back to (4.70), we can repeat the same process until the ZMP error becomes sufficiently small.

This is a very efficient algorithm. According to Nishiwaki et al. [73] it takes only 140 [ms] in generating a walking pattern for three steps (3.2 s) of the humanoid robot H7 [113] which have 32 DOF using dual Pentium II 750MHz. They have realized joystick controlled real-time walking by generating three future steps at every step cycle and by properly connecting them.

4.4.3 On-Line Walking Pattern Generation

In this section, we introduce the pattern generation method which is currently used for our humanoid robot HRP-2.

1 ZMP Tracking Control

Regarding a cart-table model as a dynamical system, one can imagine a servo system which realize a target ZMP tracking by feedback control as shown in Fig. 4.37.

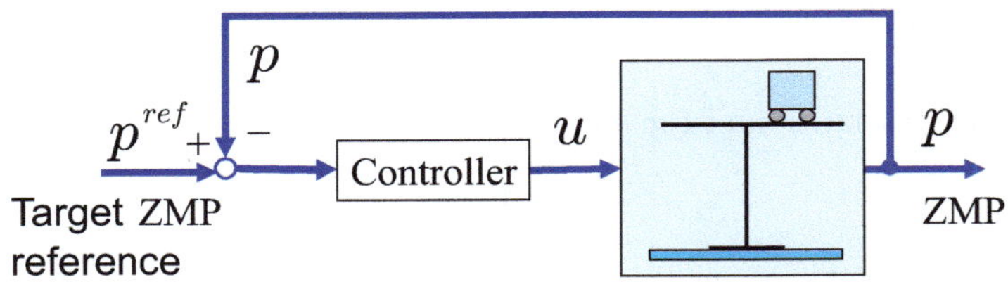

Fig. 4.37 Servo controller to track the target ZMP

We define a derivative of cart acceleration (jerk) as a system input to treat a cart-table model in a framework of the standard modern control theory,

$$u = \dddot{x}.$$

By using this input, we can rewrite the ZMP equation (4.64) into the following *state space representation*

$$\begin{aligned} \frac{d}{dt}\begin{bmatrix} x \\ \dot{x} \\ \ddot{x} \end{bmatrix} &= \begin{bmatrix} 0 & 1 & 0 \\ 0 & 0 & 1 \\ 0 & 0 & 0 \end{bmatrix}\begin{bmatrix} x \\ \dot{x} \\ \ddot{x} \end{bmatrix} + \begin{bmatrix} 0 \\ 0 \\ 1 \end{bmatrix} u \\ p &= \begin{bmatrix} 1 & 0 & -\frac{z_c}{g} \end{bmatrix}\begin{bmatrix} x \\ \dot{x} \\ \ddot{x} \end{bmatrix}. \end{aligned} \tag{4.71}$$

Starting from this equation, the modern control theory gives us a systematic way to design a controller [55]. However, we cannot obtain an appropriate walking pattern by using such controllers. For an example, suppose the walking motion of Fig. 4.38(a) which specifies a robot to walk forward one step of 30 cm. The target ZMP have step-like change at 1.5 s, but keeps constant before and after it. Notice that the CoM motion starts before the change of the ZMP. This means that the cart must move before the change of input in the system of Fig. 4.37. On the other hand, in an ordinary servo system, we have the output motion with a certain delay after a change of reference as in

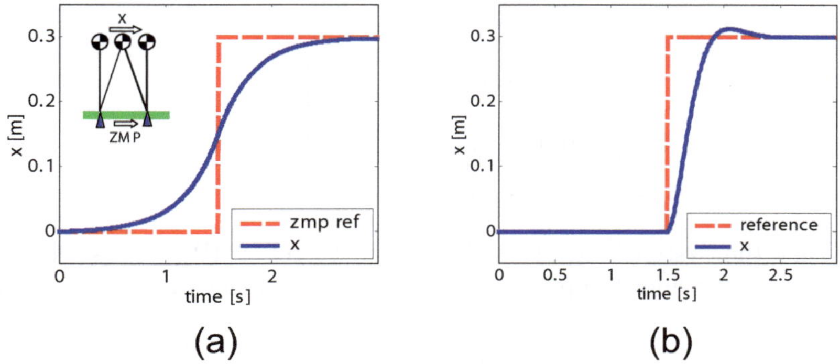

Fig. 4.38 Input-output causality (a) One step walking. ZMP (broken line), CoM (bold line). The CoM starts moving before the step change of the ZMP. (b) In an ordinal servo system, the output (bold line) moves after the change of reference input (broken line).

Fig. 4.37(b) and this is normal causality. In biped walking pattern generation, a future information must go back to and affect the past!

2 Preview Control System

Although we need future information, we do not have to develop a time-machine[15]. For example, when driving a car, we are always monitoring the road ahead and using information about the future location of the car for steering. This can be regarded as using future information for smooth driving. To appreciate the importance of using such future information, imagine driving at a high speed on a freeway with the top half of the windshield obscured, so you can only see a few meters ahead.

A method which utilizes future information is called *preview control* [128, 82, 86, 130]. Let us introduce a controller design based on this theory.

At the beginning we discretize a continuous-time system of (4.71) using a sample time of Δt to design a digital controller

$$\begin{cases} \boldsymbol{x}_{k+1} = \boldsymbol{A}\boldsymbol{x}_k + \boldsymbol{b}u_k \\ p_k = \boldsymbol{c}\boldsymbol{x}_k \end{cases} \tag{4.72}$$

where

$$\begin{aligned} \boldsymbol{x}_k &\equiv [\, x(k\Delta t) \; \dot{x}(k\Delta t) \; \ddot{x}(k\Delta t) \,]^T, \\ u_k &\equiv u(k\Delta t), \\ p_k &\equiv p(k\Delta t), \end{aligned}$$

[15] By the way, a realistic time-machine technology which can send a message to the past using particle physics is depicted in James P. Hogans's *Thrice upon a Time* (Baen Books). This is a master-piece Sci-Fi novel.

and

$$\boldsymbol{A} \equiv \begin{bmatrix} 1 & \Delta t & \Delta t^2/2 \\ 0 & 1 & \Delta t \\ 0 & 0 & 1 \end{bmatrix}, \quad \boldsymbol{b} \equiv \begin{bmatrix} \Delta t^3/6 \\ \Delta t^2/2 \\ \Delta t \end{bmatrix},$$
$$\boldsymbol{c} \equiv [\,1\ 0\ -z_c/g\,].$$

To let the system output p_k follow the target ZMP p_k^{ref} as closely as possible, we consider the problem to minimize the following performance

$$J = \sum_{j=1}^{\infty} \{Q(p_j^{ref} - p_j)^2 + Ru_j^2\}, \tag{4.73}$$

where Q, R are positive weights. This is called a *tracking control problem.* According to the preview control theory, this performance index J can be minimized by the following input which uses the future target references up to N steps [54].

$$u_k = -\boldsymbol{K}x_k + [f_1, f_2, \cdots f_N] \begin{bmatrix} p_{k+1}^{ref} \\ \vdots \\ p_{k+N}^{ref} \end{bmatrix}, \tag{4.74}$$

where

$$\begin{aligned} \boldsymbol{K} &\equiv (R + \boldsymbol{b}^T \boldsymbol{P} \boldsymbol{b})^{-1} \boldsymbol{b}^T \boldsymbol{P} \boldsymbol{A} \\ f_i &\equiv (R + \boldsymbol{b}^T \boldsymbol{P} \boldsymbol{b})^{-1} \boldsymbol{b}^T (\boldsymbol{A} - \boldsymbol{b}\boldsymbol{K})^{T*(i-1)} \boldsymbol{c}^T Q. \end{aligned} \tag{4.75}$$

The matrix $\boldsymbol{P}$ is a solution of

$$\boldsymbol{P} = \boldsymbol{A}^T \boldsymbol{P} \boldsymbol{A} + \boldsymbol{c}^T Q \boldsymbol{c} - \boldsymbol{A}^T \boldsymbol{P} \boldsymbol{b} (R + \boldsymbol{b}^T \boldsymbol{P} \boldsymbol{b})^{-1} \boldsymbol{b}^T \boldsymbol{P} \boldsymbol{A} \tag{4.76}$$

which is called a Riccati Equation[16].

By observing (4.74), we can see that a preview controller consists of a state feedback (the first term of right hand side) and a feed-forward of a inner product between the future target reference up to N steps and the weights $[f_1, \ldots, f_N]$ (the second term).

3 Improvement of Preview Controller

We observed an offset tracking error of ZMP in a long distance walking pattern generated by (4.74). To solve this problem, we rewrote (4.72) into the following expanded form:

[16] You do not have to worry about this complicated matrix equation. By using a command **dlqr** of Matlab Control System Toolbox or GNU Octave, you will immediately obtain the numerical solution of $\boldsymbol{P}$ and $\boldsymbol{K}$.

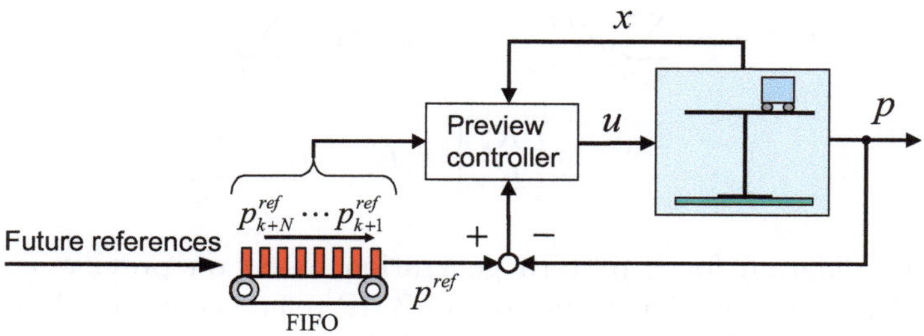

Fig. 4.39 Walking pattern generation by preview control

$$\begin{cases} \boldsymbol{x}^*_{k+1} = \tilde{\boldsymbol{A}}\boldsymbol{x}^*_k + \tilde{\boldsymbol{b}}\Delta u_k \\ p_k = \tilde{\boldsymbol{c}}\boldsymbol{x}^*_k \end{cases}, \tag{4.77}$$

where the new input and state vector were taken as

$$\Delta u_k \equiv u_k - u_{k-1}, \quad \Delta \boldsymbol{x}_k \equiv \boldsymbol{x}_k - \boldsymbol{x}_{k-1},$$
$$\boldsymbol{x}^*_k \equiv \begin{bmatrix} p_k \\ \Delta \boldsymbol{x}_k \end{bmatrix}.$$

The matrices are

$$\tilde{\boldsymbol{A}} \equiv \begin{bmatrix} 1 & \boldsymbol{cA} \\ \boldsymbol{0} & \boldsymbol{A} \end{bmatrix}, \quad \tilde{\boldsymbol{b}} \equiv \begin{bmatrix} \boldsymbol{cb} \\ \boldsymbol{b} \end{bmatrix},$$
$$\tilde{\boldsymbol{c}} \equiv [1\ 0\ 0\ 0].$$

Let us design a controller for the system of (4.77) to minimize the following performance

$$J = \sum_{j=k}^{\infty} \{Q(p^{ref}_j - p_j)^2 + R\Delta u_j^2\}. \tag{4.78}$$

The preview controller is

$$\Delta u_k = -\tilde{\boldsymbol{K}}\boldsymbol{x}^*_k + \sum_{j=1}^{N} \tilde{f}_j p^{ref}_{k+j} \tag{4.79}$$

where $\tilde{\boldsymbol{K}}, \tilde{f}_j$ are gains obtained by substituting $\tilde{\boldsymbol{A}}, \tilde{\boldsymbol{b}}, \tilde{\boldsymbol{c}}, Q$ and R into (4.75) and (4.76). By summing up equations (4.79) for $k = 1, \ldots, N$, we can obtain a preview controller for the original system of (4.72):

$$u_k = -K_s \sum_{i=0}^{k} (p_j^{ref} - p_j) - K_x \boldsymbol{x}_k + \sum_{j=1}^{N} g_j p_{k+j}^{ref} \tag{4.80}$$

$$\begin{bmatrix} K_s \\ \boldsymbol{K}_x \end{bmatrix} \equiv \tilde{\boldsymbol{K}}, \quad g_j := \sum_{i=j}^{N+j} \tilde{f}_i.$$

A block diagram for a pattern generation based on preview control is illustrated in Fig. 4.39. The future target ZMP reference is stored in a FIFO (First-In-First-Out) buffer visualized as a belt conveyer, and its output value is regarded as the *current* reference. The preview controller calculates the control input using the ZMP reference on the FIFO buffer and the state of the cart. The cart state $x, \dot{x}$ is the result of the pattern generation, the CoM motion which satisfies the target ZMP.

The CoM trajectory calculated by the proposed method and the resulted ZMP are shown in Fig. 4.40. The upper graph is the motion along the walking

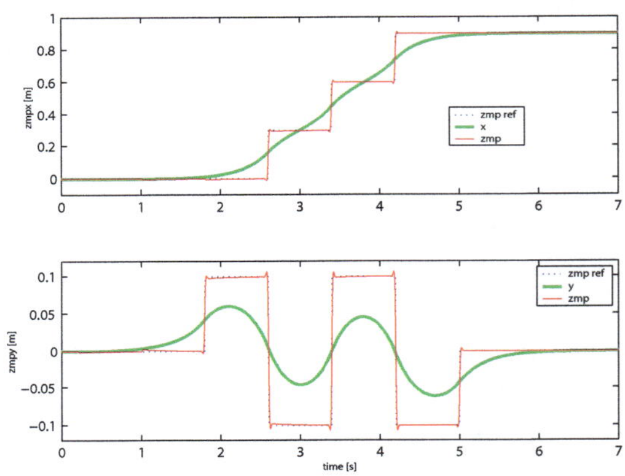

Fig. 4.40 CoM trajectory obtained by preview control

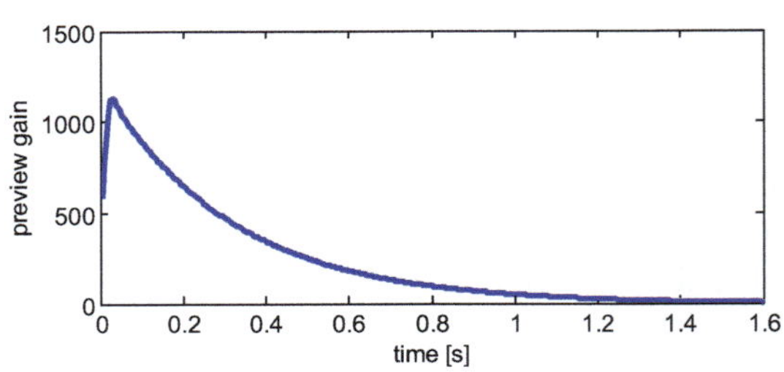

Fig. 4.41 Preview control gain g_j

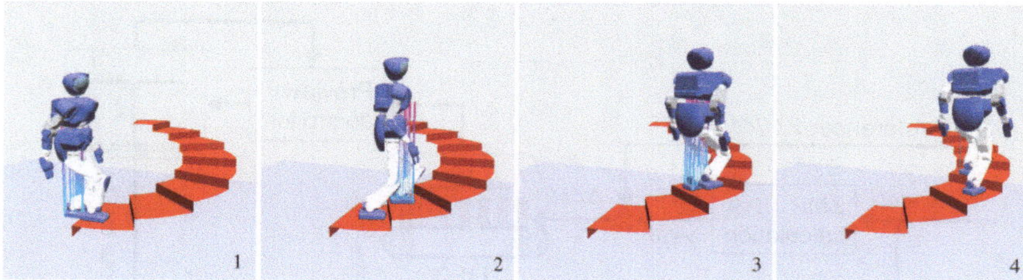

Fig. 4.42 Walk simulation on spiral stairs

direction and the lower graph is for the lateral direction[17], and we can see the proper CoM motions are generated for step-like target ZMP and pulse like target ZMP, respectively. The preview gain used for this calculation is shown in Fig. 4.41. Since the preview gain becomes negligibly small at 1.6 s in this graph, we can conclude that the information more than 1.6 s future ahead will not contribute the control performance.

Figure 4.42 shows a walking pattern on spiral stairs using the above explained method.

4.4.4 Dynamics Filter Based on Preview Control

1 Structure of a Dynamics Filter

Since a preview control based pattern generation depends on a cart-table model, it does not guarantee the stability of the motion which cannot be represented by the simplified model, for example, significantly changing upper body posture while walking. In such a case, however, we can use a cart-table model as an error system around the target motion, and can calculate the modification of the CoM trajectory to compensate for the ZMP error by using preview control again.

Figure 4.43 shows the entire structure of the proposed system. The inputs are the target ZMP (ZMP^{ref}) and the whole state of the robot*Robot state*, which consists of the angles and the speeds of all the joints, the pelvis configuration, its speed and angular velocity. We can calculate the ZMP error (ΔZMP) from these inputs, and put it into the FIFO buffer. Also put *Robot state* into another FIFO. By retrieving Δ ZMP from the FIFO output after a certain time delay, we have the future ZMP error with respect to the delayed moment on the FIFO buffer. Thus, we can calculate a proper compensation of CoM by using preview control. By applying the CoM compensation to the delayed *Robot state*, we obtain improved joint trajectories which better satisfies the target ZMP reference. Generally, a system of Fig. 4.43 is called

[17] A plane made of the heading axis and the vertical axis is called sagittal plane and a plane made of the lateral axis and the vertical axis is called lateral plane.

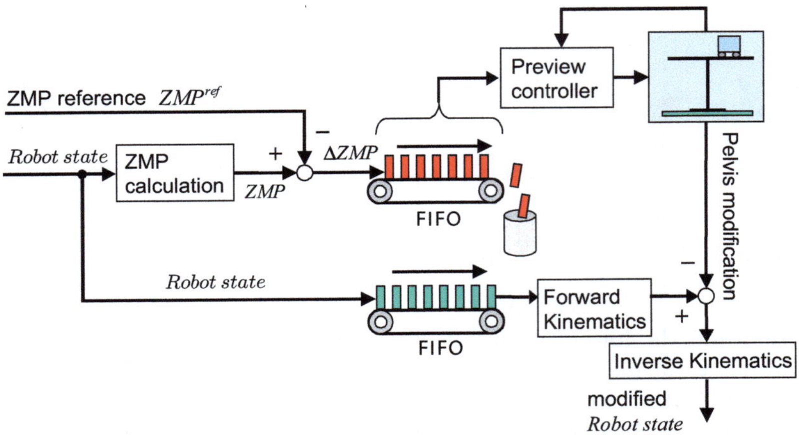

Fig. 4.43 Structure of a dynamics filter based on preview control

a *dynamics filter*. A dynamics filter converts a given motion pattern into an improved one which satisfies the desired properties [93].

2 Evaluation of the Dynamics Filter

As an example, let us make a pattern where HRP-2 performs a squat in the middle of walking and goes back to normal walking after that, as shown in Fig. 4.44. First, we simply add a squatting motion to a walking pattern assuming constant height of CoM and the resulted ZMP trajectory is shown in Fig. 4.45(a). It is observed that the ZMP (bold line) approaches the boundary of the support polygon (dotted lines) and the walking pattern has very small stability margin. Therefore, the robot falls down immediately after the squatting motion in the dynamic simulation.

This walking pattern was modified by the dynamics filter of Fig. 4.43 and we obtained the pattern of Fig. 4.45(b). We used the delay time of 0.8 s for the FIFO buffers and a preview controller without integrator (see (4.74)). The maximum absolute ZMP error which was originally 0.11 m decreased to 0.05 m by applying the dynamics filter. Since the new walking pattern has

Fig. 4.44 Example walking pattern obtained preview control based dynamics filter

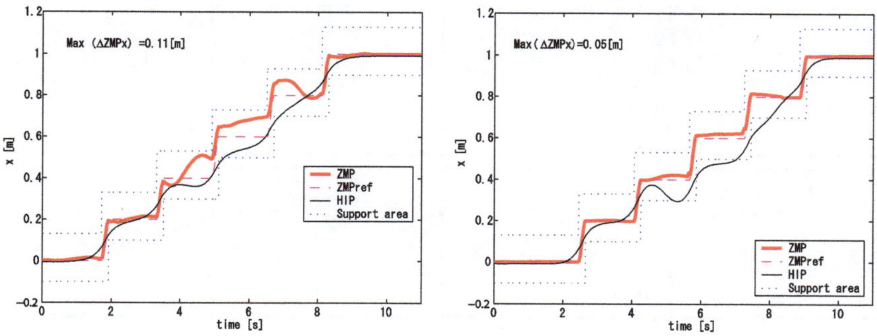

(a) Without dynamics filter (b) With dynamics filter

Fig. 4.45 Effects of the dynamics filter: ZMP(bold lines) is modified to be inside of support polygon boundary(dotted line) thus sufficient stability margin is acquired

enough stability margin, the robot could successfully walk in the dynamic simulation.

The modified walking pattern is shown in Fig. 4.44. By applying the proposed dynamics filter, we can obtain a realizable walking pattern even though the original pattern largely deviated from a simple cart-table model dynamics.

4.4.5 Advanced Pattern Generators

The method of preview control is not the only way for online walking pattern generation. As one of the practical methods, Harada et al. proposed to use an analytical solution of the ZMP equation [60]. Later, this was improved by Morisawa et al. for more efficient and responsive pattern generation [85]. These methods were used to realize HRP-2's reactive walking.

Preview control belongs the general scheme called Model Predictive Control (MPC), which calculates the control input by optimizing the future trajectory. Based on MPC, Wieber proposed a walking pattern generation which does not require a prescribed ZMP [137, 9]. By this method, ZMP and CoM trajectories can be simultaneously generated from the given profiles of the support polygon.

4.5 Stabilizer

Figure 4.46 shows snapshots of HRP-2 walking on an uneven floor. As explained in the beginning of this chapter, in a real environment which contains unmodeled errors and floor unevenness, a biped robot following a prepared pattern suffers a rapid evolution of errors between the reference and the actual state and falls down in a few steps. To suppress the error development and guarantee the walking motion along the specified walking pattern we use a stabilizer.

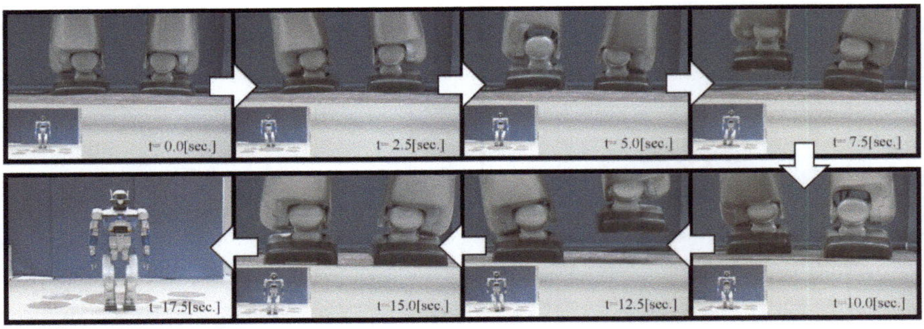

Fig. 4.46 Feet of HRP-2 walking on an uneven floor

4.5.1 Principles of Stabilizing Control

In this section, we introduce basic principles which are used to construct a stabilizer. To simplify the descriptions, we explain a method to stabilize around a state of standstill. Since a stabilizer works to absorb a small errors around the reference walking pattern, we should not lose generality with this approach.

1 Control by an Ankle Torque

We can stabilize whole body balance or posture by using the ankle torque of the support leg by modeling the whole robot as a simple inverted pendulum. An ankle torque, τ and a linear inverted pendulum have a relationship of

$$\ddot{x} = \frac{g}{z_c}x + \frac{1}{Mz_c}\tau. \tag{4.81}$$

The simplest feedback law to stabilize this pendulum is as following[18]:

$$\tau = -k_p x - k_d \dot{x}, \tag{4.82}$$

where k_p, k_d are the feedback gains determined for given response frequency ω and damping coefficient ζ,

$$k_p \equiv M(z_c\omega_n^2 + g), \quad k_d \equiv 2Mz_c\zeta\omega_n.$$

Although this control law looks easy, its implementation is difficult. First, accurate control of ankle torque is an extremely difficult problem for most walking robots equipped with high-reduction gears. Moreover, to realize

[18] For a use of more advanced control law, one must consider the problem of integrator windup, an unexpected accumulation of the integrator while the ankle torque saturates.

stable walking, we need fine tuning of feedback gains and torque limiters to suppress the toe/heel lift-off caused by excessive ankle torque.

This method was used by many biped robots developed in 1980s and 1990s, for example, WL-10RD by Takanishi et al. [10], Idaten II by Miyazaki and Arimoto [109, 32], Kenkyaku-2 by Sano and Furuhso [46] and Meltran II by Kajita and Tani [116].

2 Control by Modifying Foot Placements

The second method based on a inverted pendulum model is to modify foot placements for the stabilization. For this control, we can apply the same principle of the walking speed control explained in section 4.3.3(see Fig.4.23) on page 129). There are few robots which were stabilized by foot place modification, such as BIPER-3, a stilt walker developed by Shimoyama and Miura [41], and the series of hopping robots by Raibert and his colleagues [97].

3 ZMP Control by CoM Acceleration

Let think about a stabilization method based on a cart-table model. In this case we must measure the ZMP to design a feedback controller, and by assuming the time constant of the ZMP sensor as T, the ZMP equation is

$$p = \frac{1}{1+sT}(x - \frac{z_c}{g}\ddot{x}). \tag{4.83}$$

The state space representation having the cart acceleration $\ddot{x}$ as the input is

$$\frac{d}{dt}\begin{bmatrix} p \\ x \\ \dot{x} \end{bmatrix} = \begin{bmatrix} -1/T & 1/T & 0 \\ 0 & 0 & 1 \\ 0 & 0 & 0 \end{bmatrix} \begin{bmatrix} p \\ x \\ \dot{x} \end{bmatrix} + \begin{bmatrix} -z_c/(gT) \\ 0 \\ 1 \end{bmatrix} \ddot{x}. \tag{4.84}$$

We can design a state feedback law to stabilize this system as

$$\ddot{x} = -k_1 p - k_2 x - k_3 \dot{x}. \tag{4.85}$$

The feedback gains k_1, k_2, k_3 can be determined by a standard control theory like a pole placement or the LQ optimal control. This control law was introduced as *torso position compliance control* by Nagasaka, Inaba and Inoue [70][19]. This control law is effective for a walking robot with feet of high stiffness and was used to stabilize humanoid robots H5 and H7[113]. A simpler but effective controller is proposed by Choi, Kim and You [139].

As an alternative method, Okada, Furuta and Tomiyama proposed a stabilization method based on the CoM acceleration control via dynamic change of

[19] In the original work, they used the different procedure to derive the equivalent control law. Napoleon et al. shows another interpretation for their control law in terms of the zero-dynamics control theory [94].

sampling time and implemented to control the biped robot MK.3 and the humanoid robot morph3 [143].

4 Body Posture Control by Crotch Joints

For most walking robots, we desire that the body maintain an upright posture during walking. The easiest way is to rotate the crotch(hip) joints so that the body keeps the desired state based on a posture sensor readout. This is possible even for point contact feet since the torque around the crotch joint is generated by the friction force on the ground. This control is implemented in the Raibert's hopping robots [97], and a humanoid robot developed by Kumagai et al. [84].

5 Model ZMP Control

As a novel method of body posture control, Hirose, Takenaka et al. proposed a *model ZMP control*. According to their explanation, it works as

> *When the body of the real robot was inclined more forward than the model, the model body is more strongly accelerated than the planned trajectory. This changes the target inertial force and the target ZMP then goes more backward than the original ZMP, hence producing posture recovery in the real robot* [83, 62].

Let us consider the physical meaning of this model ZMP control using a version of cart-table model of Fig. 4.47. We assume the table foot has a free rotating joint which inclines the table at an angle of θ relative to the vertical position. The cart acceleration $\ddot{x}$ corresponds to the body acceleration.

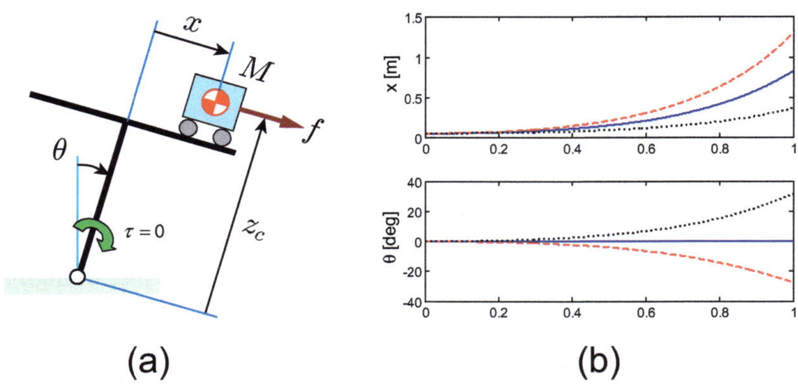

Fig. 4.47 (a) A cart-table model with free rotating joint. θ is positive in the clockwise direction. (b) Cart acceleration and the table inclination. Bold lines: with proper acceleration the table keeps upright ($\ddot{\theta} = 0$), Broken lines: with excessive acceleration the table rises ($\ddot{\theta} < 0$), Dotted lines: with insufficient acceleration the table sinks ($\ddot{\theta} > 0$). The model ZMP control uses this phenomenon.

The equation of motion obtained by Lagrange's method is

$$\begin{cases} (x^2 + z_c^2)\ddot{\theta} + \ddot{x}z_c - g(z_c \sin\theta + x\cos\theta) + 2x\dot{x}\dot{\theta} = \tau/M \\ \ddot{x} + \ddot{z}_c - \dot{\theta}^2 x + g\sin\theta = f/M \end{cases}, \tag{4.86}$$

where τ is the torque acting at the table foot and f is the force to accelerate the cart on the table. By linearizing the first equation around $\theta, \dot{\theta} = 0$ and substituting $\tau = 0$, we obtain

$$(x^2 + z_c^2)\ddot{\theta} = gx + gz_c\theta - z_c\ddot{x}. \tag{4.87}$$

Next, assume the target position of the cart x_d is generated by the dynamics of

$$\ddot{x}_d = \frac{g}{z_c}(x_d - p_d), \tag{4.88}$$

where p_d is the target ZMP. Intuitively speaking, when the target ZMP is placed backword ($p_d < 0$) the cart accelerate strongly and when it is placed forward ($p_d > 0$) the cart acceleration becomes small. Since (4.88) is just a model to generate the acceleration, the target ZMP can be assigned at the outside of the support polygon.

Now, suppose we can control the cart acceleration as we want. By substituting $x = x_d$ and (4.88) into (4.87) we get

$$\ddot{\theta} = \frac{gz_c}{x_d^2 + z_c^2}\theta + \frac{g}{x_d^2 + z_c^2}p_d. \tag{4.89}$$

From this result, we can see that the table inclination θ can be controlled by the target ZMP p_d, and its dynamics is determined by the cart position and gravity acceleration.

6 Impact Absorption by Joint Backdrive

Embedding a joint torque sensor and applying the reference position as a function of the torque measurement, we can realize a virtual spring-damper system. Takanishi et al. utilized such system to absorb the vibration at touch-down impact [10]. Also when a joint has a small reduction ratio (1/1 to 1/50 approximately), it backdrives by the external forces from the environment thus a position control system behaves as a spring-damper by itself. Kenkyaku 1, a biped robot developed by Furusho et al. utilized this for its walking control [47]. Sorao, Murakami and Ohnishi reported an impact absorption at support exchange using an impedance controller which depends on a sensor-less torque measurement by *disturbance observer* [75].

7 Stabilization by LQ Control

A walking robot can be modeled as a multi-input-multi-output (MIMO) system having joint torques $\boldsymbol{u}$ as its input vector and output as the state of all links. Let us define a state vector $\boldsymbol{x}$ consisting of all the posture and speed variables of all the links, and linearize the motion equation from $\boldsymbol{u}$ to $\boldsymbol{x}$. By a straightforward application of LQ control theory, we get a state feedback law of

$$\boldsymbol{u} = -\boldsymbol{K}\boldsymbol{x},$$

where $\boldsymbol{K}$ is a feedback gain which is a huge $N \times 2N$ matrix for the number of joint N. This approach was applied to a 6 DOF biped, CW-2 by Mita et al. [133] and 12 DOF biped developed by Yoshino [145]. Especially, the latter work reports that the robot could successfully walk at 3 [km/h] on the floor with unevenness of 6 [mm].

4.5.2 Stabilizing Control of Honda Humanoid Robot

A practical stabilizing control system is constructed by combining multiple control principles described above. Let review one such successful implementation.

Fig. 4.48 Humanoid robot P2 (1996) (By courtesy of Honda Motor Co., Ltd.)

Figure 4.48 shows a humanoid robot P2 developed by Honda Motor Co., Ltd. and officially announced in 1996. Its control system has a state-of-art quality even at the time of this printing.[20] and its technology is well disclosed [83, 62]. The walking control system of P2 is illustrated in Fig. 4.49. The feedback path from the body inclination to the ground reaction force corresponds the ankle torque control described in 1. It should be noted that the passive compliant elements inserted in the feet of P2 makes this control easier

[20] It is assumed that the same type of controllers are also used for another Honda humanoids P3 and ASIMO which are the successors of P2.

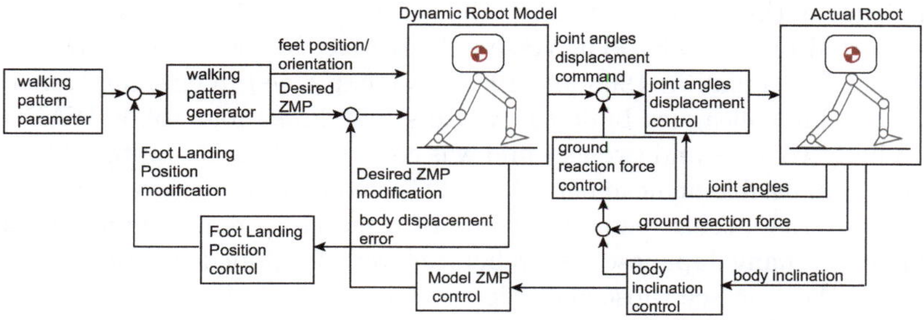

Fig. 4.49 The walking control system of Honda P2

to implement. When the robot body gets an excessive inclination the model ZMP control **5** works to resolve it. The horizontal body displacement error caused by this is corrected by the foot landing position control **2**. As we see, a robust walking control of P2 is supported by an ingenious combination of multiple control schemes.

4.5.3 Advanced Stabilizers

Recently, we proposed a new stabilizer based on a state space model of LIPM with ZMP control lag [117]. This stabilizer allows our new humanoid robot HRP-4C to walk on a certain level of uneven ground as well as to perform human-like walking with toe supporting [69]. One of the key issues of this stabilizer is a proper setting of the state feedback gain which was discussed by Sugihara [125]. Another important issue is distribution of the desired ZMP. A related concept of stabilization based on a total ground force distribution was discussed by Hyon [42].

A huge disturbance during walking, for example a kick on the body, can immediately cause the robot to deviate from a preplanned trajectory and consequently fall down. The robot can avoid falling even in this case, if it can re-plan the walking pattern by letting the deviated state be the initial condition. By properly implementing this concept, very robust walking can be realized [72, 85, 51].

As we discussed in Section 4.5.1, adaptive change of foot placement is a basic strategy for stabilization. Pratt et al. defined the **capture point** which is a point on the ground where the robot can step to achieve its complete stop [50]. Based on the capture point concept, Englsberger et al. proposed a simple framework for real-time walking pattern generation and stabilization [44].

4.6 Pioneers of Dynamic Biped Walking Technology

Let us look back in the research history of dynamic biped walking control. Since the dawn of the robotics research, biped walk was widely recognized as a challenge of the highest difficulty. The earliest research to develop hardware were started by Ichiro Kato in Waseda University in 1966, and by D.C. Witt in Oxford University in 1968 [68, 19]. As we explained in chapter 1, in 1973, Kato and his colleagues built the world's first humanoid robot WABOT-1 which had two arms and two legs and was controlled by a computer. Although it was a remarkable achievement, WABOT-1 could only perform static walking. About 1980, there was a big trend of research to realize dynamic biped walking and many Japanese researchers actively developed theories and robot hardware. We can see those activities in *Special Issue: Biped Walking Robot*, the journal of Robotics Society of Japan, Vol.1, No.3. In 1986, there already exists many biped robots which could perform dynamic walking as shown in Fig. 4.50 [27].

In the same year, a group of researchers in Honda Motor Co.,Ltd. started a secret project of biped walking robot. After complete silence for 10 years[21], their efforts have borne fruit as the humanoid robot P2. The humanoid robot P2 suddenly appeared with an excellent hardware and walk control technology, and it depressed researches who have been working on biped walking control until that time. At the same time, it also inspired many researchers by demonstrating the great possibility of humanoid robots. This yielded the power to start the *Humanoid Robotics Project* (1998-2002), the national R&D project of Japan.

People who get used to watch ASIMO or QRIO might think the robots in Fig. 4.50 are primitive. However, the researchers who developed them by struggling with poor computers and weak actuators of these days are the real pioneers of biped walking technology. Indeed, most of the modeling and control technique for biped walking were developed until those days. Moreover, I can hardly believe that Honda's research project was independent from those many research works already published in 1986. The progress of science and technology is not done by a genius in one night. There is no brilliant breakthrough independent from the heritage. Therefore, the true progress can be done only by the open exchange of ideas in a collaboration of many researchers, beyond institutes, beyond countries.

4.7 Additional Methods for Biped Control

In this section, we introduce biped robots and controls based on methods totally different from the concepts so far explained.

[21] Compared with universities, it is estimated that more than 100 times the financial budget as well as manpower was invested by Honda during this time. Since the robotics community held a mostly negative view on biped robotics in those days, the long-range vision of the Honda people should be highly respected!

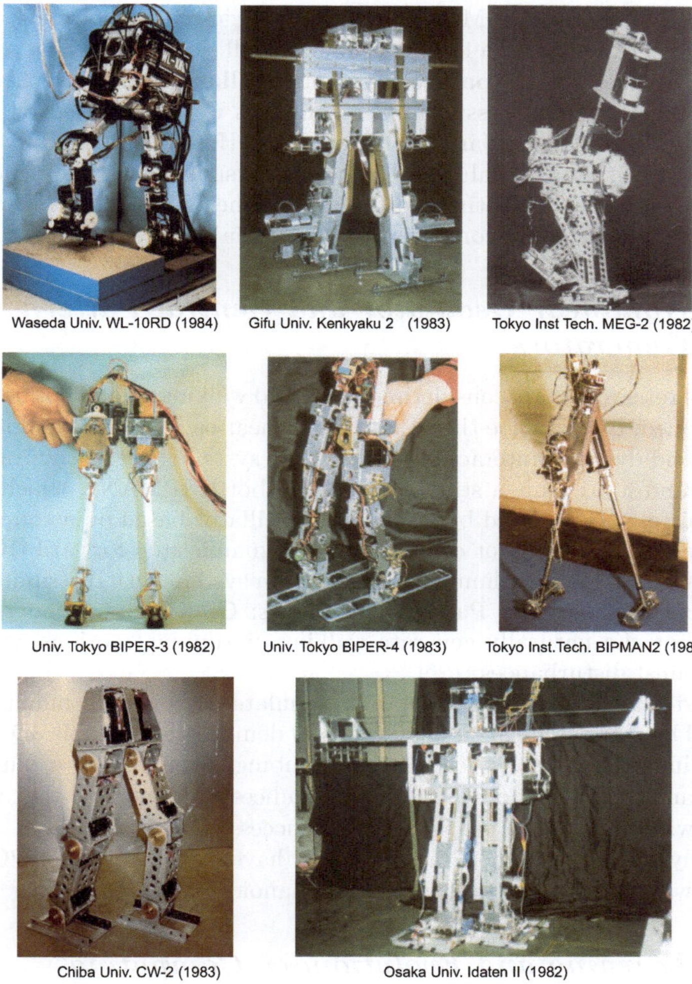

Fig. 4.50 Biped robots which could walk dynamically before 1986 [27]

4.7.1 Passive Dynamic Walk

Passive dynamic walkers are mechanisms which can walk down a shallow slope only using potential energy. Tad McGeer intensively analyzed its dynamics and designed a 2D walker with free knee joints which can perform stunningly human like walking without any actuation or control [131, 132]. Following and improving this concept, a 3D passive walker was also realized by Collins, Wisse and Ruina [111].

Passive dynamic walking has attracted many researchers who are seeking for the fundamental nature of biped locomotion. There are works analyzing its chaotic behavior [14, 74] and its robust stability [141].

In the original paper of McGeer [131], he suggested a "powered" passive walker, which can walk on a level ground or uphill slopes by adding small actuators. It is called a **semi-passive dynamic walker**. There exist theoretical works [7, 29, 91] and successful implementations were developed by Hobbelen and Wisse [21] and by Narioka, Tsugawa and Hosoda [71]. Collins, Ruina, Tedrake and Wisse have evaluated the power consumption of their successful semi-passive walkers and estimated that one of their robots is approximately ten times more efficient in comparison with Honda's ASIMO [110].

4.7.2 Nonlinear Oscillator and Central Pattern Generators

A group of researchers are considering that biped walking should not be analytically planned, but must be the result of nonlinear oscillations emerging from feedback and dynamic interaction between the system and the environment.

Katoh and Mori built a stilt type biped robot, BIPMAN2, which used a stable limit cycle generated by a nonlinear oscillator based on a coupled van der Pol equation. The robot could take one dynamic step forward [100].

Taga et al. simulated a human's muscle-born system with distributed nonlinear oscillators (Central Pattern Generators: CPGs) and found the simulated robot can naturally generate walking or running motions which are robust against disturbances [126].

Inspired from Taga's work, Hase et al. simulated detailed 3D human model controlled by a hierarchical CPG system and demonstrated stable 3D walking and running[61]. They also used a hill-climbing algorithm to optimize the CPG parameters for various performance indices and reported that walking patterns with human-like properties were successfully obtained.

Recently, Hyon, Morimoto and Kawato have demonstrated CPG-based dynamic walking by their human-sized humanoid robot [107].

4.7.3 Learning and Evolutionary Computing

The most radical approach might be to build a robot which can learn or evolve walking by itself. Doya built a simple biped robot consists of three links, and a self learning system of hardware in the loop. The system generated a walking pattern randomly and learned by using a hill-climbing algorithm using actual traveled distance as its performance index [57]. As the result, the robot could acquire a variety of walking patterns including a jumping gait and a tumbling gait which were not expected. de Garis designed a neural network based walking control system whose weights are tuned by a genetic algorithm, and simulated a evolution of biped walking [20].

Tedrake et al. designed a 3D semi-passive walker equipped with four actuators whose motion can be acquired by online reinforcement learning. It is reported that the robot could learn an adequate walking pattern for the various floor conditions within about twenty minutes [106].

Chapter 5
Generation of Whole Body Motion Patterns

In this Chapter we will explain how you generate whole body motion patterns for a humanoid robot. There is a lot of ongoing research in this area and the ultimate method is still yet to be created. So therefore we will simply introduce examples that are being used today.

5.1 How to Generate Whole Body Motion

Let us consider the motion shown in figure 5.1. The Robot has to first move from a standing posture to a kneeling posture, lift the luggage from the ground and carry the luggage 1[m] forward. How would you go about generating Whole Body Motion like this? Intelligent readers would probably have a plan formed by now.

To generate a walk pattern you can use the methods discussed in the previous chapter. The problem lies in the way you actually go about generating the motion to enable the robot to pick up the box. Why can't you simply use the same method as used by industrial robot manipulators?

Industrial 6 axes robot manipulators usually come with a device called a teaching pendant as shown in figure 5.3. This device enables you to set the initial position and attitude of the end effector, which may be for welding or painting or grasping. Furthermore, you can set way points through which the end effector must pass, and the controller will interpolate between these points and generate control angle trajectories. This is possible because a 6 axes industrial robot manipulator has the minimum amount of joints that enable it to have 6 degrees of freedom. So if you set the position and attitude which amounts to 6 degrees of freedom you can solve the joint angles using inverse kinematics.

The major difference with a robot with many degrees of freedom such as a humanoid robot, is that even if you set the position and attitude of the end effector you will not get a unique solution, there will be several. You can try this out for yourself by grasping a door handle standing up and then squatting

S. Kajita et al., *Introduction to Humanoid Robotics*, 159
Springer Tracts in Advanced Robotics 101,
DOI: 10.1007/978-3-642-54536-8_5, © Springer-Verlag Berlin Heidelberg 2014

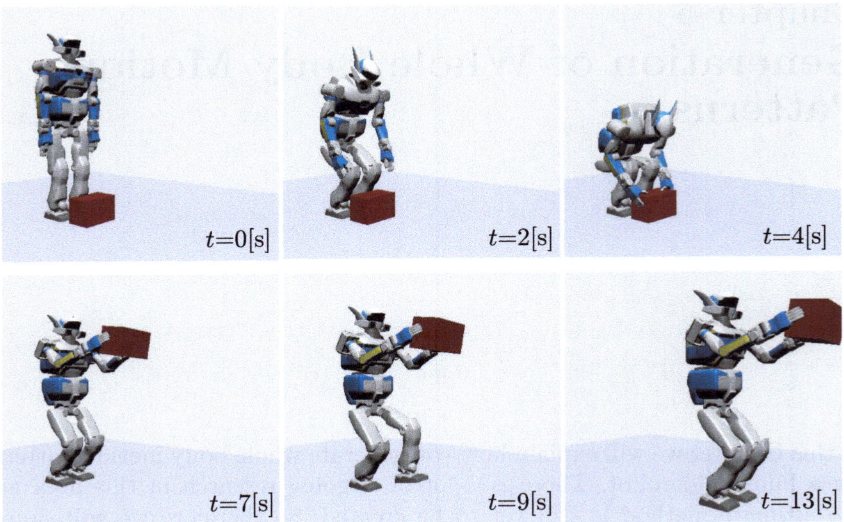

Fig. 5.1 Simple Task for the HRP-2

down. A further complication is the fact that a humanoid robot is not bolted down to the floor like an industrial robot manipulator. This means you need to not only make sure that the initial posture, the intermediate postures and the goal posture will enable the robot to keep it's balance, you must be sure that the ZMP trajectory stays within the support polygon which is defined by the robot's contact points with the ground.

Due to this, you will see that the majority of the methods discussed below will first generate a rough full body motion and the modify it so that the ZMP trajectory stays within the support polygon. Figure 5.4 shows the steps that are required to generate full body motion. In step A we first generate whole body motion to fit the requirements. In step B we make the necessary adjustments to enable the robot to balance. Most humanoid robots have sensor feedback loops that keep the robot stable so the actual motion that is used is a further modification of the trajectory generated in step B[1].

5.2 Generating Rough Whole Body Motion

There are around three methodologies that have been proposed for generating whole body motion that roughly suits your requirements.

1. Use Motion Capture and get data from a human being.

[1] If the motion created in step B is perfect step C may not be an absolute necessity. However, as you probably are aware, the world is not a perfect place so it is not realistic to think that you can pre-define all the parameters related to the real world.

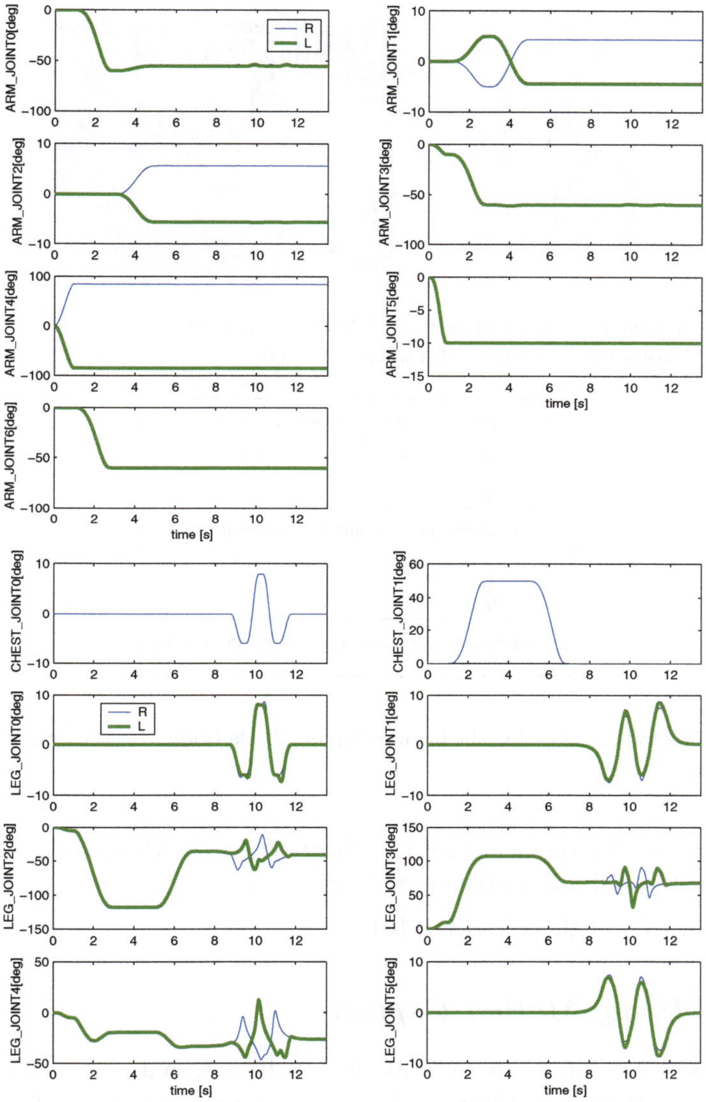

Fig. 5.2 Time Graph of the major joints on the HRP-2. First 7[s] are lifting the box, 7 to 13[s] is the walking motion.

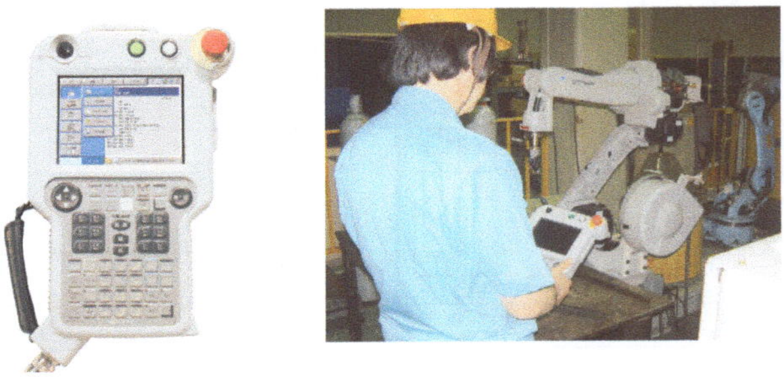

Fig. 5.3 A Teaching Pendant(left) and Actual Use(right)
(Photograph courtesy of Yaskawa Electric)

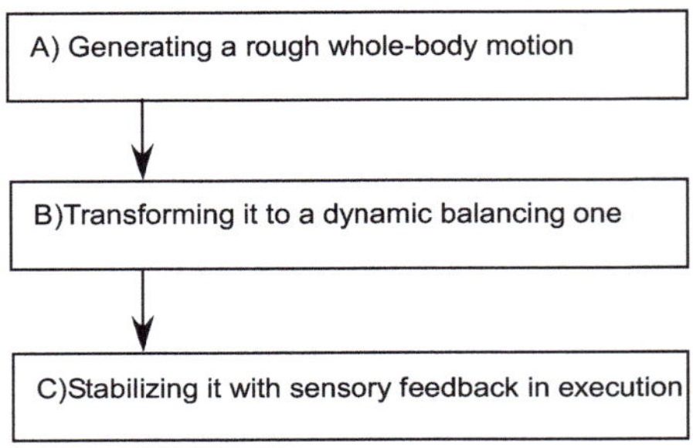

Fig. 5.4 Flowchart Showing the Generation of Whole Body Motion for a Humanoid Robot

2. Create a GUI(Graphical User Interface) or use a CAD system.
3. Use high-speed multivariate search methods.

We will go over these in detail below.

5.2.1 Using Motion Capture

Humanoid robots are made to be like human beings, so it is a natural step to try to use data captured from real human motion. To actually capture this data, you usually simply use a Motion Capture System[2]. Fig 5.5 shows motion capture data of a person dancing the Tsugaru Jongara-bushi(a dance native to northern Japan)[121]. This enables you to capture whole body motion from a human being rather easily. However, there are subtle differences between a humanoid robot and an actual human being. This means you cannot simply send this data to a robot and expect it to move properly.

[2] A system which can record the trajectory of a point is space that has been marked using specialised markers that the system is tuned to detect. This information is digitized and stored in a computer.

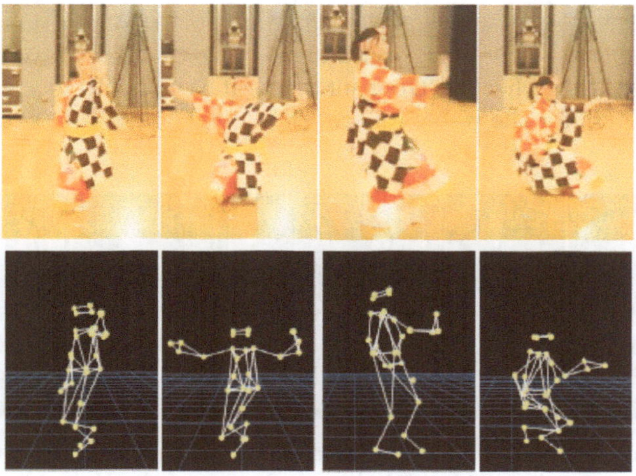

Fig. 5.5 Sample Data from Motion Capture of a Person Doing the Tsugaru Jongara-bushi Dance[121]

5.2.2 Using a Graphical User Interface

In computer generated animation, you see people or humanized animals and robots, move around as if they exist in the real world. In this sense, tools that are used for computer graphics are also valid for generating motion for humanoid robots. A method that has been proposed for generating humanoid robot motion, is the pin/drag interface [76].

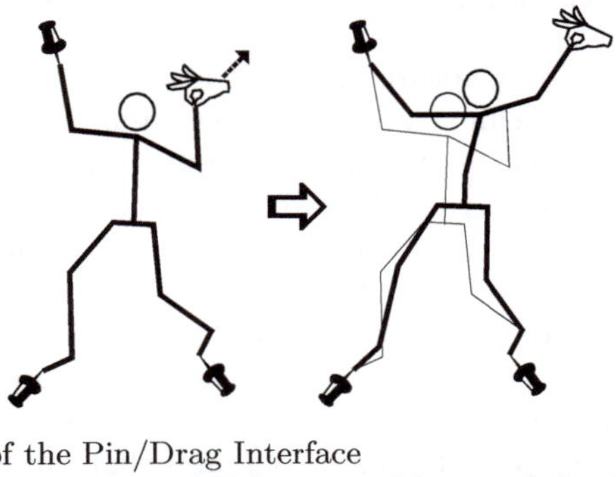

Fig. 5.6 Concept of the Pin/Drag Interface
(Photograph courtesy of Nakamura - Yamane Lab., University of Tokyo)

The basic concept is shown in fig 5.6. You *pin* the links that should not move and move the end points you want to move to wherever you want them to move. By using the Pin/Drag interface, it is possible to create natural looking[3] motion for the whole robot.

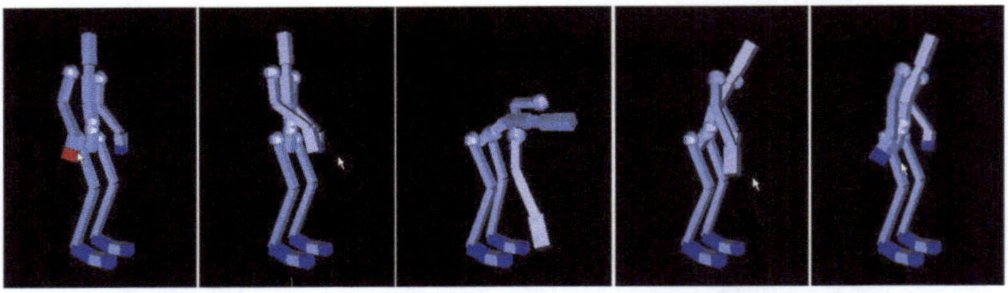

Fig. 5.7 Sample Full Body Motion Generated using the Pin/Drag Interface (Photograph courtesy of Nakamura-Yamane Lab., University of Tokyo)

Fig 5.7 shows motion that was generated with the toes and heels of both legs and the left hand pinned down. The right hand was them moved down and then up, by dragging the part in the screen, over a period of four seconds. You can see that you get a natural looking motion of the robot picking something up from the floor. However, this motion only satisfies the physical restraints that you sent during motion generation. It does not satisfy dynamic restraints such as ZMP. If you use this motion as it is, there is still a possibility that the robot will fall.

5.2.3 Using High Speed Multivariate Search Methods

While the Use of Motion Capture and GUI based methods may be intuitive they are still cumbersome and can be time consuming. The third solution is to use methods such as RRT[48] which do a high speed multivariate search. By searching the vector space created by all the joints you can generate motion that satisfies conditions such as reaching under a table without colliding with obstacles such as the surface of the table[49].

In this method, by preparing 3D data of the Humanoid Robot and it's surrounding environment you can generate whole body motion such as reaching under a chair which is shown in fig 5.8, in under 6 seconds, automatically. With this method, you search for stable postures that satisfy the physical constraints and then create joint trajectories that connect the postures smoothly. Due to this fact, you cannot generate dynamically stable whole body motion.

[3] You actually use solutions that make the second order summation minimum.

By courtesy of James Kuffner (The Robotics Institute, CMU/ Digital Human Research Center, AIST)

Fig. 5.8 Sample Whole Body Motion of Robot Reaching Under Chair, Generated using RRT Search

5.3 Converting Whole Body Motion Patterns to Dynamically Stable Motion

Motion created using the methods introduced in section 5.2 are not dynamically stable so you cannot simply use them to control a real robot. So the next step becomes the conversion of the generated trajectories from a series of statically stable postures to dynamically stable joint trajectories, as shown in fig 5.4. In this section we introduce, the Dynamics Filter, the Auto-balancer and the Strict Trunk Motion Computation. The Dynamics Filtering Method using Preview Control which we introduced in the previous section is also one way to convert trajectories to dynamically stable ones.

5.3.1 Dynamics Filter

The Dynamics Filtering Method is a filter that converts physically impossible whole body motion to a physically feasible one [79][76]. This method consists of a Controller and an Optimizer(Fig 5.9). The Controller takes the raw untested trajectory θ_G^{ref} and the current state of the robot θ_G, $\dot{\theta}_G$ as input. It consists of local and global feedback loops that calculate the next goal joint acceleration $\ddot{\theta}_G^d$. This goal joint acceleration is not necessarily something that can be achieved at this point. The Optimizer looks for a solution that satisfies physical constraints on the robot(floor drag, direction of force from floor contact points), and also keeps the amount of acceleration required at the joints $\ddot{\theta}_G$, minimum.

In Fig. 5.10 top half, we show motion capture data. The bottom half shows data that has been converted using the Dynamics Filter. You can see that the positioning of the feet have been converted to more realistic trajectories.

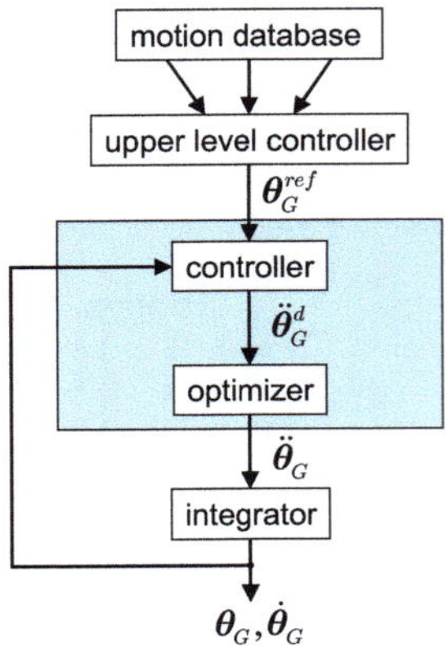

Fig. 5.9 Concept of the Dynamics Filter[79]

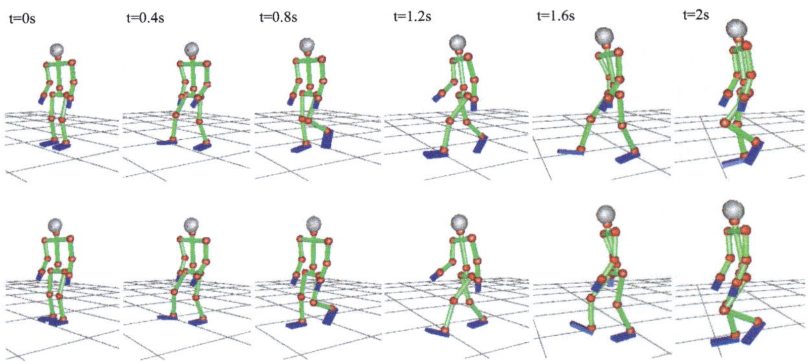

Fig. 5.10 Conversion using the Dynamics Filter
(Photograph Courtesy of Nakamura-Yamane Lab., University of Tokyo)

5.3.2 Auto Balancer

The Auto-Balancer[4] performed second order optimization methods to joint angles at each timestep[147]. In this method, emphasis was on static stability, so although it cannot be used to stabilize walking, it was a valid for stabilizing motion in which the robot remained in the same place and the support

[4] This name was originally used in a comic called "Patlabor"(Shogakukan) by Yuki Masami. The system that enabled the robots(called labor's in the comic) to keep their balance as called the Auto-Balancer. Tamiya. et. al. who were inspired by this system used the same name for their system.

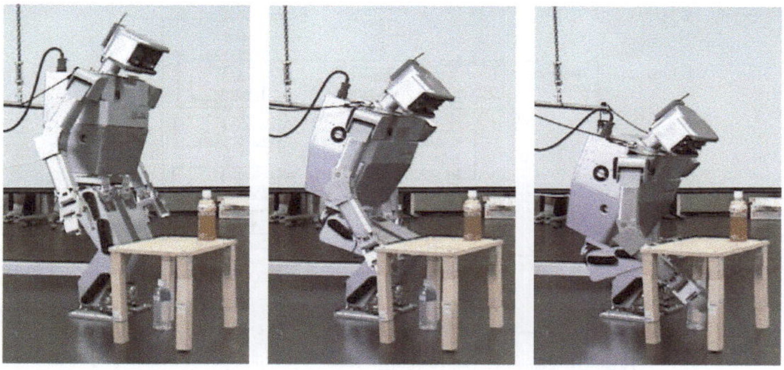

Fig. 5.11 Conversion Using an Auto-Balancer
(Photograph Courtesy of JSK Lab. University of Tokyo)

polygon did not change[5]. The Auto-Balancer enables the robot to keep it's balance by doing the following:

1. Keep the center of the gravity of the robot along an axis that goes through a given point in the support polygon.
2. Keep the moment around this axis within a given amount.

The amount of moment that is allowed around the axis is calculated from the conditions that you must satisfy to keep the ZMP point within the support polygon, and the condition you must satisfy to keep the moment force below zero until the next time frame. This method of balance compensation is basically a second order operations research problem that solves for whole body motion that remains as close to the original requirement as possible. The Auto-Balancer performs these calculations for all the time steps.

Fig 5.11 shows a case where whole body motion was generated using the RRT method shown in fig 5.8 and then stabilized using the Auto-Balancer and then actually performed by the Humanoid Robot H6[49].

5.3.3 Strict Trunk Motion Computation Algorithm

Trunk Motion Compensation Algorithm creates a trajectory for the body base taking as input the trajectories of the legs, ZMP and the hands. The calculated result is use as the whole body motion trajectory [53].

Fig 5.12 shows the flowchart used in this algorithm. The assumptions used in the linearization and collision detection and avoidance are that, the height of the waist joint is not changed, the simplified model of the arms only move within the horizontal plane and that the approximated center of gravity of the upper body is not to be moved along the Z axis. By performing a Fast

---

[5] It is possible to apply this method to walking but it is only valid for doing slow static walk[112].

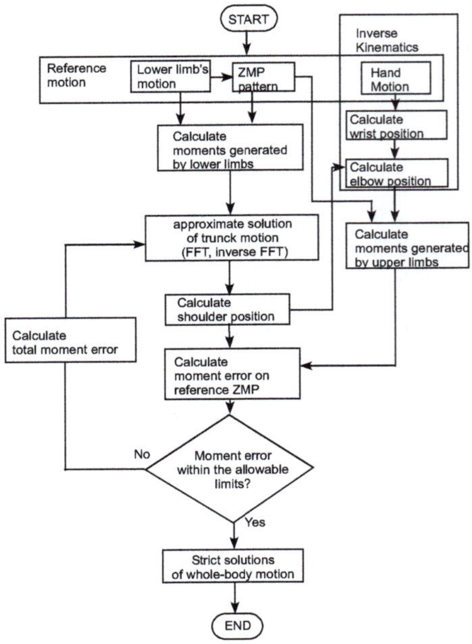

Fig. 5.12 Flowchart to compute the Strict Trunk Motion [53]

Fig. 5.13 Humanoid Robot WABIAN
(Photograph Courtesy of Takanishi Lab, Waseda University)

Fourier Transform(FFT) on the 3 axis moment around the goal ZMP point and then performing a inverse FFT transform you can get a good estimate for the trunk motion. By using this on the actual model of the robot you can calculate the amount of moment error around the goal ZMP point that occurred due to the fact that an approximation was used instead of the real model. By adding and integrating these values you can re-calculated a closer estimation. By iterating over this whole procedure you can make the solution converge to an analytical solution.

This algorithm allows you to get an accurate analytical solution but the downside is that it requires looping over a lot of calculations. So therefore, a method which make an assumption of the actual error that occurs from the approximation model based on the moment error from the nth calculation. By using this method algorithms have been proposed that reduce the amount of calculation that is required by up to 1/100.

To deal with non recurring motion such as the start and end of a motion, you can create data that has a long period of time with no motion before and after the actual motion and apply the same algorithm. The Strict Trunk Motion Computation [53] is used to generate whole body motion for the Humanoid Robot WABIAN shown in fig 5.13.

5.4 Remote Operation of Humanoid Robots with Whole Body Motion Generation

When making Humanoid Robots dance or play instruments you have enough time to prepare the data so you can use offline motion generation methods. However if you are using the Humanoid Robot at a location you have never been to before, or in a new situation you need to generate whole body motion online, or on the fly in real-time.

Sadly the robot's awareness of it's own environment and planning capabilities are no way near as high as we humans. So the methods that have been proposed try to use this capability that human being have inherently as much as possible.

Generating motion for humanoid robots is the same in the case for online as it was offline as shown in fig 5.4. You first generate a rough approximation (step A), then you modify it to enable the robot to balance properly (step B) and after that modify again to compensate for dynamic motion (step C). But things change a lot when you have to do all this on the fly and in real-time. However rough they may be, creating whole body motion trajectories in real-time is not an easy task. So when you are doing remote operation of a humanoid robot, you modify steps A, B above and do the following. In step A you generate partial body motion trajectories and then in step B you create whole body motion trajectories that enable the robot to keep stable, and then in step C you use a stabilizer which keeps the robot stable using sensor feedback and modify the whole body motion trajectories.

Below we go over how you go about doing remote operation using whole body motion with humanoid robots.

For step A we will look at “Remote Generation of Whole Body Motion using the Operation Point Switching Method”. Step B will focus on “Whole Body Motion Generation using Split Momentum Control”. We have already discussed a method which you can use to stabilize the robot using sensor feedback in the previous chapter.

5.4.1 Remote Generation of Whole Body Motion Using the Operation Point Switching Method

As humanoid robots have large degrees of freedom, to actually create whole body motion in real-time you need a large system like a Motion Capture System. However, humanoid robots aren’t exactly like humans so captured data cannot be used without modification.

So let us reflect over how exactly we move our bodies daily. People focus on different areas depending on what they want to do. They switch their attention to the part that is most important to the operation in hand and consciously move it. At the same time, the joints that are non-essential to the operation are moved “unconsciously” to either keep your balance or enable easier operation of the task at hand.

For instance, when reaching for a bottle on the table, your attention will be on the hand you are going to use to pick the bottle up. When you are about to sit down, you will focus your attention on your bottom while lowering your body towards the chair. When you are going to kick a ball, you focus your attention on the foot you are going to use to kick the ball(fig 5.14).

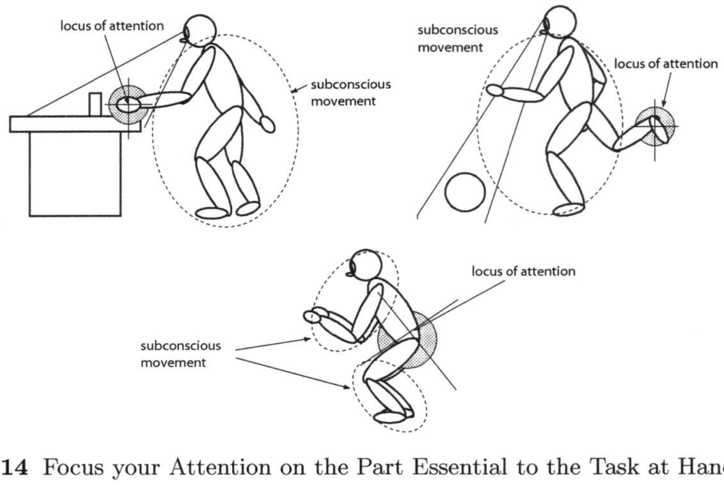

Fig. 5.14 Focus your Attention on the Part Essential to the Task at Hand

The part we focus our attention on is usually moved in a given restricted amount of degrees of freedom, so it is usually possible to use a simple input device such as a joystick to indicate the required motion. On the other hand, the parts that are moved unconsciously can be implemented as part of the control that enables the robot to stay balanced. This enables you to control the robot without having to think about the geometric and dynamic differences between humans and robots. This is the basis of the Attention Point Switching Method [25].

As an example we will explain the way you would use two joysticks to generate full body motion for a humanoid robot. First we allocate functions to the eight buttons on the joysticks as shown in fig 5.15. The functions will be indication of the head, right/left hands, right/left wrists, the body and the right/left feet, and the switching between the three coordinates, world, body and local. The operator presses the button that represents the position he want to operate and uses the levers to specify direction to generate full body motion that changes the attitude and position of the body part of the humanoid robot that is being operated.

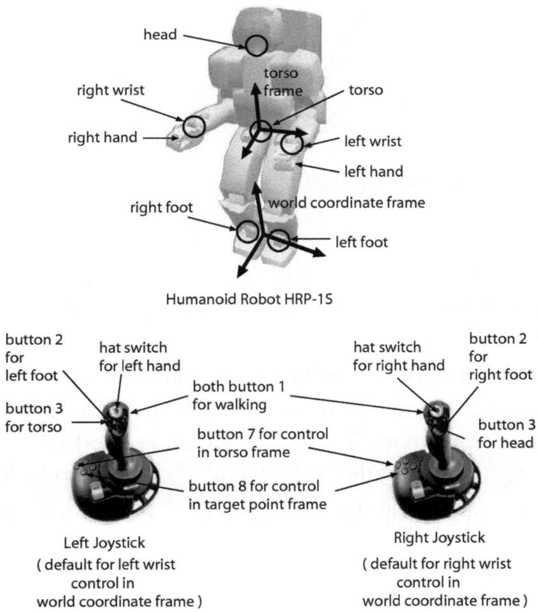

Fig. 5.15 Example Button Allocation to Joystick for Allocation of Humanoid Robot Body Part and Operational Coordinates

5.4.2 Full Body Motion Generation of Stable Motion Using Split Momentum Control

When you have knowledge of accurate physical parameters of the Humanoid Robot, you can generate full body motion for the robot by specifying the amount of change in translational and angular momentum and using this as a basis for calculating the joint angle speeds. This is called Split Momentum Control [108].

First you model the humanoid robot using a six degree of freedom body link that has four open links, which is shown in 5.16.

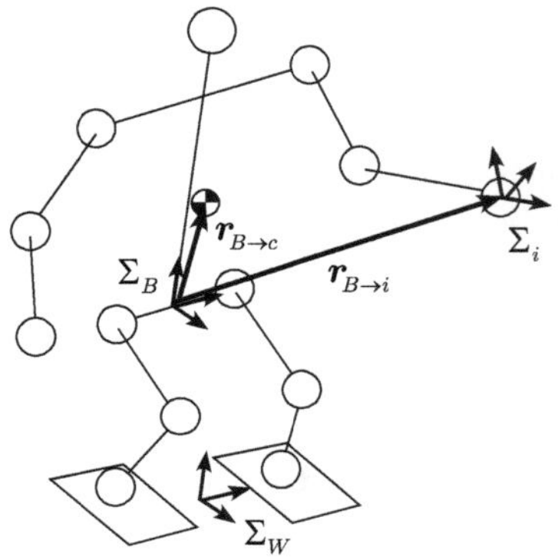

Fig. 5.16 Model of Humanoid Robot

By doing this you can use the following equations to calculate the goal speeds $\boldsymbol{v}_i^{ref}$, goal angular velocities $\boldsymbol{\omega}_i^{ref}$ body coordinates Σ_B velocity $\boldsymbol{v}_B^{trg}$, angular velocity $\boldsymbol{\omega}_B^{trg}$

When given the the goal velocity of the end of the limbs $\boldsymbol{v}_i^{ref}$ and goal angular velocity of the end of the limbs $\boldsymbol{\omega}_i^{ref}$ in limb end coordinates Σ_i, defined within the world coordinate Σ_W which is fixed to the floor, and the target velocity $\boldsymbol{v}_B^{trg}$ and angular velocity $\boldsymbol{\omega}_B^{trg}$ in the body local coordinate Σ_B, the goal joint angles of the ith limb, $\dot{\boldsymbol{\theta}}_i^{ref}$ can be calculated using the following.

$$\dot{\boldsymbol{\theta}}_i^{ref} = \boldsymbol{J}_i^{-1}\left\{\begin{bmatrix}\boldsymbol{v}_i^{ref}\\ \boldsymbol{\omega}_i^{ref}\end{bmatrix} - \begin{pmatrix}\boldsymbol{E} & -\widehat{\boldsymbol{r}}_{B\to i}\\ 0 & \boldsymbol{E}\end{pmatrix}\begin{bmatrix}\boldsymbol{v}_B^{trg}\\ \boldsymbol{\omega}_B^{trg}\end{bmatrix}\right\} \tag{5.1}$$

Here, $\mathbf{J}_i^{-1}$ is the general inverse matrix of the Jacobian of the ith limb. $\boldsymbol{E}$ is a 3×3 unit vector, $\boldsymbol{r}_{B\to i}$ is the vector which starts from the origin of the body coordinates and ends at the origin of the operational coordinate Σ_i.

The operator ^ is the skew symmetric matrix which is equivalent to the cross product.

The $\boldsymbol{v}_i^{ref}$ and $\boldsymbol{\omega}_i^{ref}$ in eq. 5.1 is defined by the remote operation to be done. For example, if the operator wants to make a translational movement with the wrist the operator uses the levers of the remote controller to openly specify the exact amount of movement to set to the translational velocity vector $\boldsymbol{v}_1^{ref}$. During this operation all the elements of the rotational velocity vector of the right hand, $\boldsymbol{\omega}_1^{ref}$ are set to zero together with the rotational and translational velocity vectors of all the other limbs, $\boldsymbol{v}_i^{ref}$, $\boldsymbol{\omega}_i^{ref}$[6].

$\boldsymbol{v}_B^{trg}$ and $\boldsymbol{\omega}_B^{trg}$ can be calculated using split momentum control using the following equations. They rely on the humanoid robot's goal translational momentum $\mathcal{P}^{ref}$, goal angular momentum around it's center of gravity $\mathcal{L}^{ref}$ and $\boldsymbol{v}_B^{ref}$, $\boldsymbol{\omega}_B^{ref}$ which are calculated by autonomously by the robot or specified by the remote operator.

$$\begin{bmatrix} \boldsymbol{v}_B^{trg} \\ \boldsymbol{\omega}_B^{trg} \end{bmatrix} = \boldsymbol{A}^{\dagger}\boldsymbol{S}\left\{ \begin{bmatrix} \mathcal{P}^{ref} \\ \mathcal{L}^{ref} \end{bmatrix} - \sum_{i=1}^{4} \begin{pmatrix} \boldsymbol{M}_i \\ \boldsymbol{H}_i \end{pmatrix} \boldsymbol{J}_i^{-1} \begin{bmatrix} \boldsymbol{v}_i^{ref} \\ \boldsymbol{\omega}_i^{ref} \end{bmatrix} \right\} + (\boldsymbol{E}_6 - \boldsymbol{A}^{\dagger}\boldsymbol{A}) \begin{bmatrix} \boldsymbol{v}_B^{ref} \\ \boldsymbol{\omega}_B^{ref} \end{bmatrix} \tag{5.2}$$

C

$$\boldsymbol{A} \equiv \boldsymbol{S}\left\{ \begin{pmatrix} M\boldsymbol{E} & -M\widehat{\boldsymbol{r}}_{B\to c} \\ 0 & \boldsymbol{I} \end{pmatrix} - \sum_{i=1}^{4} \begin{pmatrix} \boldsymbol{M}_i \\ \boldsymbol{H}_i \end{pmatrix} \boldsymbol{J}_i^{-1} \begin{pmatrix} \boldsymbol{E} & -\widehat{\boldsymbol{r}}_{B\to i} \\ 0 & \boldsymbol{E} \end{pmatrix} \right\}$$

$\boldsymbol{M}_i$ and $\boldsymbol{H}_i$ are inertia matrices that indicates the effect induced by the total translational momentum and angular momentum of the robot to the ith link. M is the robot's total mass and $\boldsymbol{I}$ is the inertia tensor around the robot's center of gravity, $\boldsymbol{r}_{B\to C}$ is a potential vector based in the robot's body coordinate system Σ_B. $\boldsymbol{E}_6$ is a 6×6 unit matrix and † denotes a pseudo inverse matrix. $\boldsymbol{S}$, which gives you to the moment element that you want to control, is a $n \times 6$ selection matrix $(0 \le n \le 6)$ that is described as follows:

$$\boldsymbol{S} \equiv \begin{bmatrix} \boldsymbol{e}_{S_1} & \dots & \boldsymbol{e}_{S_n} \end{bmatrix}^T$$

Here, $\boldsymbol{e}_{S_i}$ is a 6×1 column vector which has the elements that correspond to the moment elements you want to control set to 1 and the rest to 0[7].

---

[6] Within the split momentum control framework it is possible to move the remaining limbs to satisfy the required motion. However, as this usually leads to unpredictable behaviour from the robot it is common to keep the remaining limbs still.

[7] If you set this to control all moment elements $A^{\dagger}$ will not have an area set to zero so $\boldsymbol{v}_B^{ref}$ and $\boldsymbol{\omega}_B^{ref}$ cannot be controlled by equations 5.1 and 5.2.

In eq. 5.2, $\boldsymbol{v}_I^{Ref}$ and $\boldsymbol{\omega}_I^{Ref}$ are specified by the remote operator. $\mathcal{P}^{Ref}$, $\mathcal{L}^{Ref}$, $\boldsymbol{v}_B^{Ref}$ and $\boldsymbol{\omega}_B^{Ref}$ can be autonomously calculated by the robot's control system. For instance, $\mathcal{P}^{Ref}$ and $\mathcal{L}^{Ref}$ can be used to keep the robot balanced. If we define a coordinate system Σ_F as being parallel to the World coordinate system Σ_W with it's origin at the center of the supporting polygon of the robot, the relationship between the total translational momentum $\mathcal{P}$ and the translational velocity in the Σ_F coordinate system $\dot{\boldsymbol{r}}_{F\to c}$ can be described as follows:

$$\mathcal{P} = M\dot{\boldsymbol{r}}_{F\to c} \tag{5.3}$$

Therefore by calculating the the goal translational momentum $\mathcal{P}^{ref}$ using the next equation, the center of gravity $\boldsymbol{r}_{F\to c}$ can be moved to a given position $\boldsymbol{r}_{F\to c}^{ref}$.

$$\mathcal{P}^{ref} = Mk(\boldsymbol{r}_{F\to c}^{ref} - \boldsymbol{r}_{F\to c}) \tag{5.4}$$

The k above is a preset gain. By adding equation 5.4 to the autonomous controller within the robot, the operator can specify motion while the robot keeps it's balance without the operator having to think about the difference between kinematics and dynamics[8].

5.4.3 Application and Experiments with the Humanoid Robot HRP-2

Fig 5.17 shows a sample implementation of the remote whole body operation systems we explained in the previous section, on the humanoid robot HRP-2. The control system mainly consists of the sub-systems, *Input Device Server*, *Whole Body Motion Generator* and the *Stabilizer* [26].

The *Input Device Server* generates motion for part of the humanoid robot using inputs based on the operation point switching method of remote robot operation.. This is implemented on a PC running Linux. The user generates motion using two three degree of freedom joysticks and the buttons on them, to set the target translation and rotational velocity of the body coordinate system Σ_B and the link being operated Σ_i.

The *Whole Body Motion Generator* uses the split momentum method to generate stable whole body motion for the robot. It is placed within the robot itself. Motion generated by this component is sent to the I/O control boards in the PC, and then onto the servo driver boards.

[8] For accuracy, using eq. 5.4 alone will only guarantee that the robot's center of gravity will stay within the support polygon. But you can get the robot to keep it's balance if you also take into consideration the goal angular momentum $\mathcal{L}^{ref}$ around the robot's center of mass and keeping it to zero, or by keeping the goal acceleration of the body coordinate system $(\dot{\boldsymbol{v}}_B^{ref}, \dot{\boldsymbol{\omega}}_B^{ref})$ at sufficiently small values.

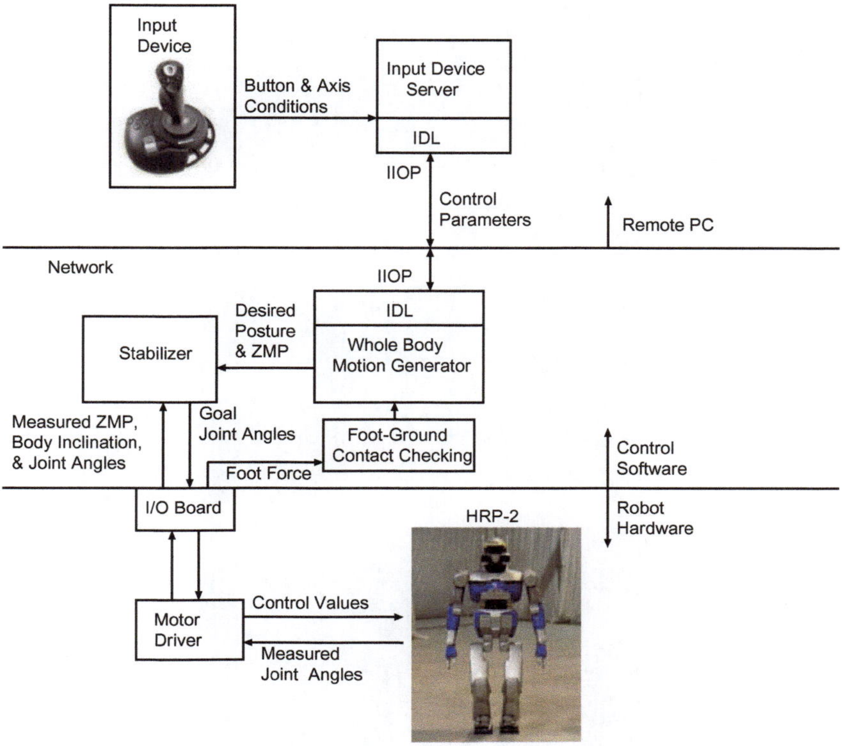

Fig. 5.17 Humanoid Robot Whole Body Remote Operation System. Uses the standardized distributed object system, CORBA to connect input devices.

The *Stabilizer* uses sensor feedback to keep the robot in a stable state.

Fig 5.18 shows images of the HRP-2 being used to pick up a can which is placed in a different position on the table and throw it away into a rubbish bin using remote operation.

The actual flow of this operation is as follows:

1. The Humanoid Robot HRP-2 is standing around 3[m] away from the table.
2. The operator first uses the joystick to point the head of the HRP-2 towards the table to locate the position of the can.
3. When the can is located the robot approaches the can. The operator indicates the direction and speed of the walk using the joysticks.
4. The robot stops and uses a 3D vision system[146] to locate the position of the can and measures the distance from the can.
5. Based on the detected location data, the robot calculates the trajectory of the hand and automatically grasps the can.
6. The operator uses remote operation to lift the can and checks that the can has been grasped properly.

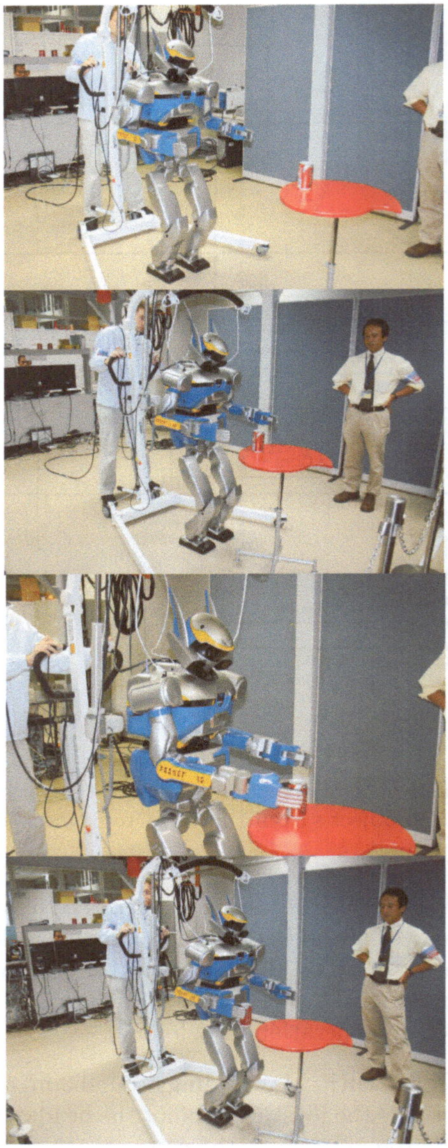

Fig. 5.18 A Remote Operation Experiment

7. The robot starts walking again and goes to the rubbish bin and throws the can away. This part is done completely by remote operation.

At one point the robot grasped the can too tightly and the can got stuck to it's hand. However the operator was able to shake the robot's hand remotely and therefore was able to drop the can. One of the benefits of implementing remote operation is the fact that you get this kind of high level error recovery automatically.

5.5 Reducing the Impact of a Humanoid Robot Falling Backwards

Anything that is not anchored to the ground can fall over. As shown in fig. 5.19 an object which, like a humanoid robot, has a high center of mass and a small supporting polygon becomes statically unstable even when tipped over a small angle.

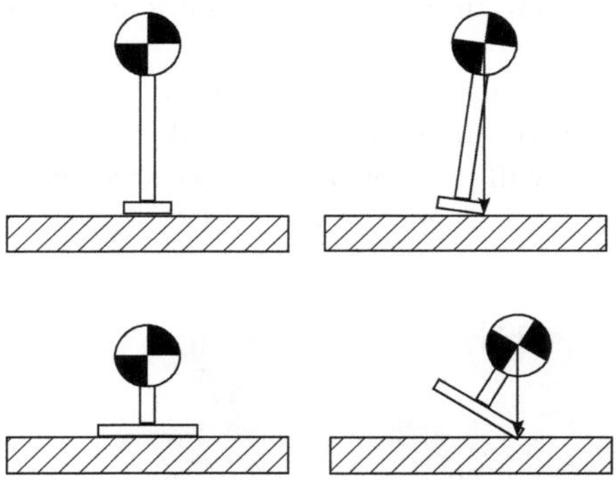

Fig. 5.19 Relation Between Static Stability and Hight of Center of Mass and Supporting Polygon Size

Until this point in the book, everything was about how to keep a robot stable to prevent it from falling down. As you can see from the fact that even human beings fall over, In reality it is not possible to prevent robots from falling over altogether. Therefore it is important that humanoid robots are able to fall over gracefully to prevent themselves from being damaged too seriously. An example of Impact Reduction Control is the one used on the HRP-2P [58].

The HRP-2P(fig. 5.20) is the prototype model of the HRP-2. The height of this humanoid robot is 158cm and the weight is 58kg [66]. On the hip of the HRP-2P are pads to absorb the shock[9]. So it is best to hit the ground

[9] In the experiment with the robot falling backwards, to improve the impact absorption of the upper body we added padding to the robot's back as well.

with as low velocity as possible. First we divided the fall itself into five states and the robots motion was controlled differently in each state.

1. SQUATTING State This is the initial state of falling where the robot's center of mass deviates too far from the supporting polygon. All control is stopped and the fall control takes over. It starts to bend the robot's knees and also starts to bend the neck, waist and arms so that the robot falls on its hips.
2. EXTEND1 State The fall proceeds. When the angle made by the ground and the line that connects the point the robot's heels touch the ground and the hip landing point θ(Fig. 5.21) goes below a certain value, the robot starts to extend it's knees to decrease the velocity at which the hips hit the ground. This also makes sure that the robot hits the ground with it's hips.
3. TOUCHDOWN State The fall proceeds to the next stage. When angle θ goes below another value, the robot prepares for landing and stops servo control of the joints.
4. EXTEND2 State When a certain amount of time elapses after the fall, the robot extends it's legs to prevent rolling on it's head due to the momentum of the fall.
5. FINISH State After enough time has elapsed and it is detected that the robot is in a stable state, the joints are extended to prepare for the standing up motion which we will go over in the next section.

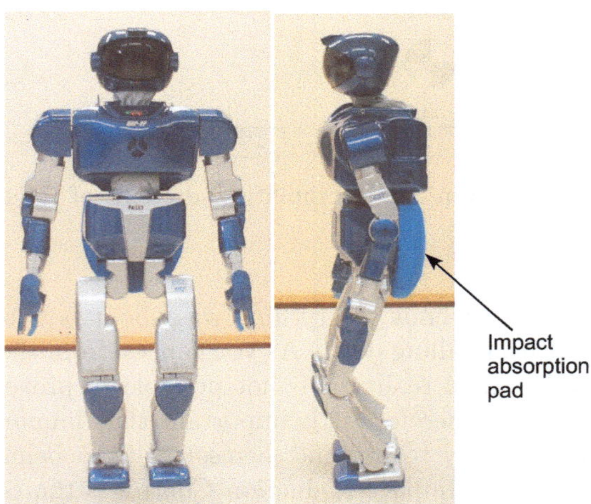

Fig. 5.20 The Humanoid Robot HRP-2P

By performing this action and reducing the amount of impact the HRP-2P was able to stand up after falling down as shown in fig. 5.22. The standing motion will be explained in the next section.

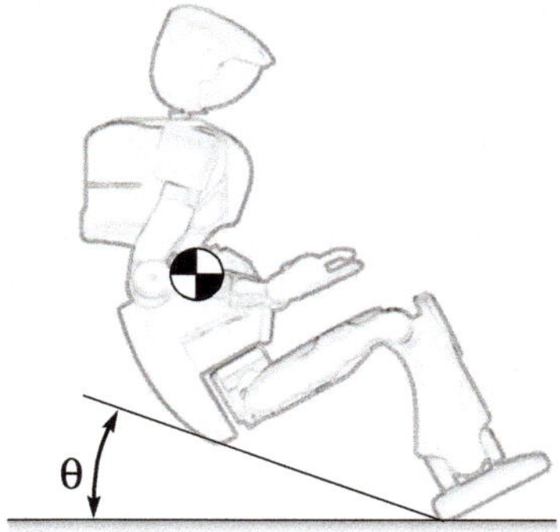

Fig. 5.21 Definition of Falling Angle θ

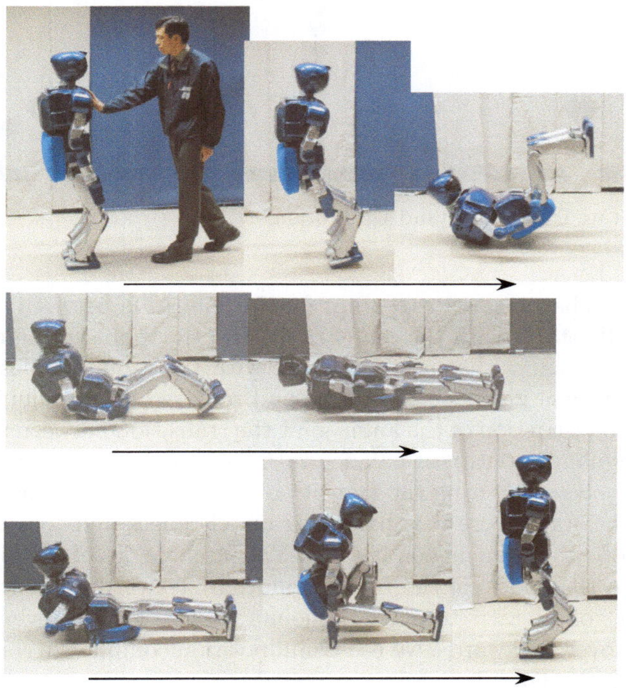

Fig. 5.22 Falling Over and Standing Back Up

5.6 Making a Humanoid Robot Get Up Again

In the previous section we explained how to reduce the impact of a fall so that the humanoid robot can prevent itself from getting damaged too seriously. What you need next is to be able to get up again. To do this you need to be able to get up from the state where the robot is lying on it's front or it's back[10]. To get up you must keep the robot's balance but when moving to a standing state you also need to disturb the robot's balance as well [28]. So to stand up you need to divide it into individual motions as shown in fig. 5.23.

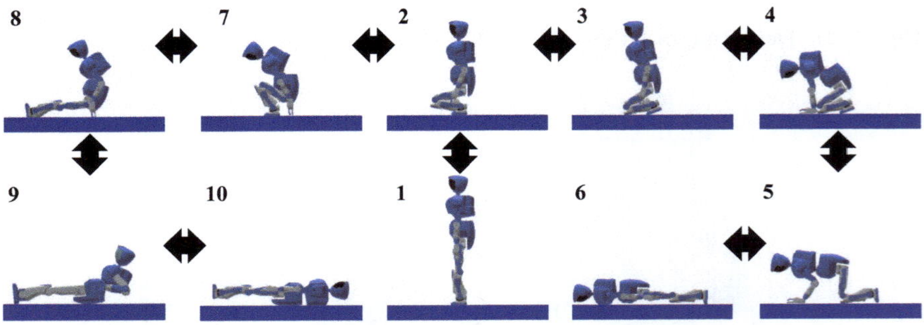

Fig. 5.23 State Chart for Stand Up Motion

All switches between states, except for the transition between states 2 and 3 are statically stable with the center of mass of the robot within the support polygon of the points that are in contact with the ground. These statically stable motions can be generated offline and replayed when required.

Due to the restriction of the ankle joint the transition between states 2 and 3 requires dynamic control of balance with the center of mass outside of the supporting polygon. During the transition from state 3 to 2, the robot must shift from the state there it is balancing on it's knees and toes to the state where the robot is balancing on it's feet. The pitch angles of the hip joints are controlled so that body of the robot is moved backwards. When rotation is detected the robot starts to move the joints to state 2. To prevent the robot from falling over backwards, we use trunk position compliance control [70] to keep the robot balanced.

We applied this control to the Humanoid Robot HRP-2P. In figs. 5.24 and 5.25 we show the robot getting up from the state where it is lying face down and also face up. The $t =$ in this figure shows the amount of time that has

[10] In the previous section we only explained how to make the robot land on it's back. However it is important that the robot be able to land on it's front. In this case the final position would be the state where the robot is lying on it's front.

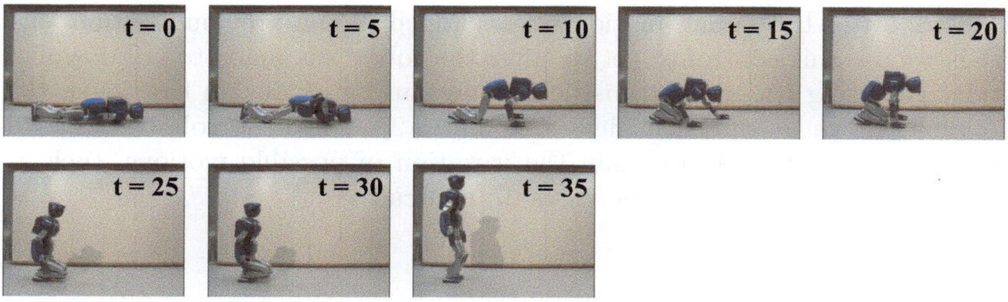

Fig. 5.24 Standing Up Motion with HRP-2 Lying Face down [28]

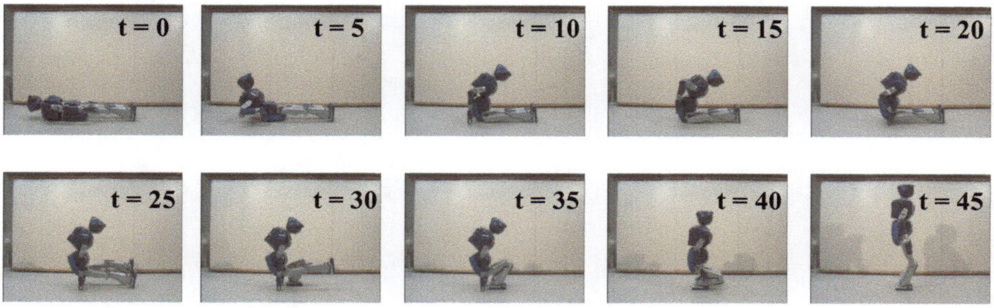

Fig. 5.25 Standing Up Motion with HRP-2 Lying Face up [28]

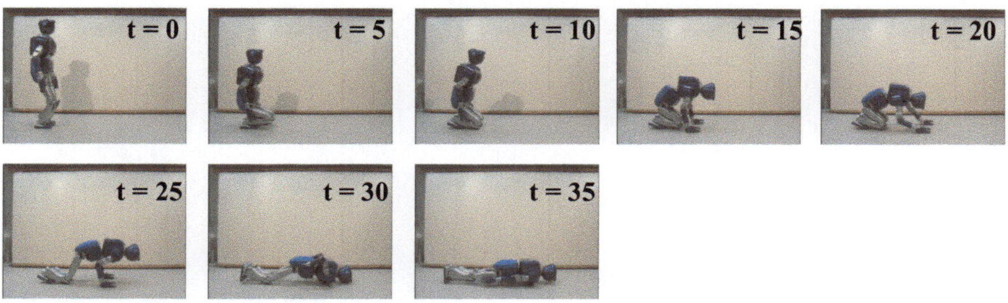

Fig. 5.26 Making HRP-2 Lie Prone [28]

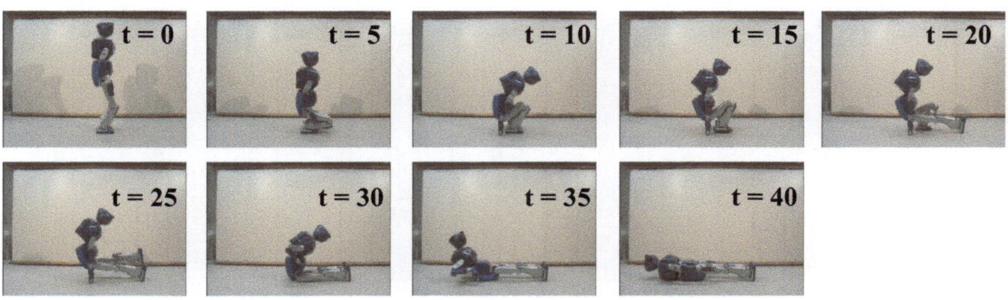

Fig. 5.27 Making HRP-2 Lie Face Up [28]

elapsed. After the falling motion we explained in the previous section (fig. 5.22) we use this control to get up after the robot was made to fall down.

By reversing the state transitions shown in fig. 5.23 you can make the robot lie down as shown in figs. 5.26 and 5.27. Once the robot is able to lie down and get back up again the variation of possible motions, such as crawling underneath a car, increases dramatically.

Chapter 6
Dynamic Simulation

Figure 6.1 shows a falling experiment of the humanoid robot HRP-2P. By performing a **dynamic simulation**, we can predict when and how strong the robot will hit the floor before actually performing such an experiment. It is also a powerful tool to develop a falling control method to minimize the landing impact and to develop durable hardware.

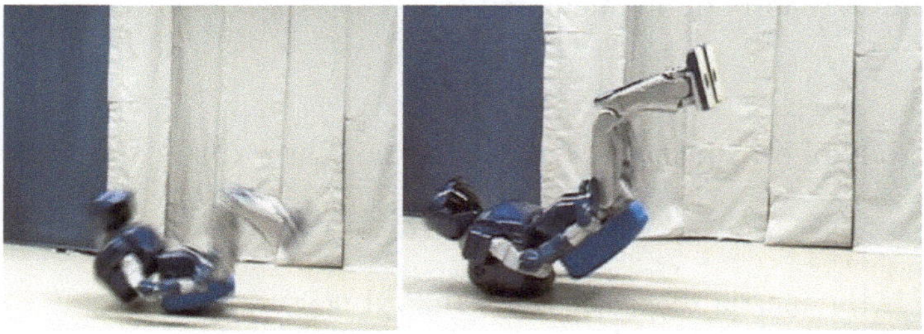

Fig. 6.1 Falling and landing experiment of HRP-2P[58]

The goal of this chapter is to show the theory and the practice of dynamic simulation. First, we will derive the equation of motion for a rotating object in zero-G space. By developing a simulator for this problem, we develop a good intuitive understanding of dynamic simulation. In the following sections, we will develop a complete rigid body simulator by introducing translation, gravity and contact forces with environment. As one of the more impressive 3D dynamic motions, a spinning top is simulated by our technique.

A dynamic simulation of a robot can be formulated as multiple rigid bodies connected by joints. At the end of this chapter, we overview an efficient algorithm developed by Featherstone.

S. Kajita et al., *Introduction to Humanoid Robotics*, Springer Tracts in Advanced Robotics 101,
DOI: 10.1007/978-3-642-54536-8_6, © Springer-Verlag Berlin Heidelberg 2014

6.1 Dynamics of Rotating Rigid Body

We first derive the equation of motion of a rigid body rotating around its center of mass, then explain a simulation program based on the equation. In this section, we assume that the CoM of the rigid body is stationary at the point where the origin of the world frame is placed.

6.1.1 Euler's Equation of Motion

Let us recall the basics of dynamics we learned in Chapter 3. The momentum $\mathcal{P}$ of an object changes by the applied force vector $\boldsymbol{f}$:$\dot{\mathcal{P}} = \boldsymbol{f}$. By substituting the definition of momentum $\mathcal{P} = m\dot{\boldsymbol{p}}$, we obtain the Newton's equation of motion

$$\boldsymbol{f} = \frac{d}{dt}\mathcal{P} \quad \Rightarrow \quad \boldsymbol{f} = m\ddot{\boldsymbol{p}}.$$

Likewise, the angular momentum $\mathcal{L}$ of an object changes by the applied moment

$$\boldsymbol{\tau} = \frac{d}{dt}\mathcal{L}. \tag{6.1}$$

With given angular velocity vector $\boldsymbol{\omega}$ and inertia tensor $\boldsymbol{I}$, an angular momentum of a rigid body is calculated as

$$\mathcal{L} = \boldsymbol{I}\boldsymbol{\omega}. \tag{6.2}$$

A rigid body of orientation $\boldsymbol{R}$ has the inertia tensor of

$$\boldsymbol{I} = \boldsymbol{R}\bar{\boldsymbol{I}}\boldsymbol{R}^T, \tag{6.3}$$

where $\bar{\boldsymbol{I}}$ is the inertia tensor at the standard posture.

Those are what we have learned in Chapter 3.

Now we can derive the equation of rigid body's rotation. By substituting (6.2) and (6.3) into (6.1), we get

$$\begin{aligned}\boldsymbol{\tau} &= \frac{d}{dt}(\boldsymbol{R}\bar{\boldsymbol{I}}\boldsymbol{R}^T\boldsymbol{\omega})\\ &= \dot{\boldsymbol{R}}\bar{\boldsymbol{I}}\boldsymbol{R}^T\boldsymbol{\omega} + \boldsymbol{R}\bar{\boldsymbol{I}}\dot{\boldsymbol{R}}^T\boldsymbol{\omega} + \boldsymbol{R}\bar{\boldsymbol{I}}\boldsymbol{R}^T\dot{\boldsymbol{\omega}}.\end{aligned}$$

By using the basic equation of rotation $\dot{\boldsymbol{R}} = \widehat{\boldsymbol{\omega}}\boldsymbol{R}$,

$$\boldsymbol{\tau} = \boldsymbol{\omega} \times (\boldsymbol{R}\bar{\boldsymbol{I}}\boldsymbol{R}^T\boldsymbol{\omega}) + (\boldsymbol{R}\bar{\boldsymbol{I}}\boldsymbol{R}^T)\dot{\boldsymbol{\omega}}.$$

Finally, by replacing the inertia tensor using (6.3), we get the following equation called Euler's equation

$$\boldsymbol{\tau} = \boldsymbol{I}\dot{\boldsymbol{\omega}} + \boldsymbol{\omega} \times \boldsymbol{I}\boldsymbol{\omega}. \tag{6.4}$$

6.1.2 Simulation of Rigid Body Rotation

Euler's equation is a bit more complicated than Newton's one. To understand its nature, let us perform a simple simulation.

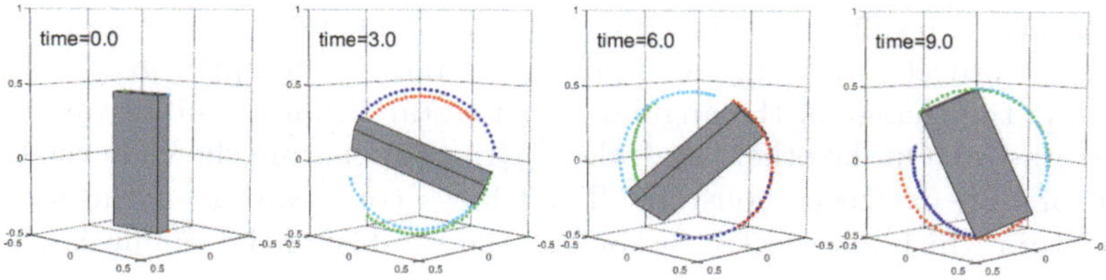

Fig. 6.2 Free rotating rigid body in zero-G: In general, free rotation in 3D does not occur around a fixed axis

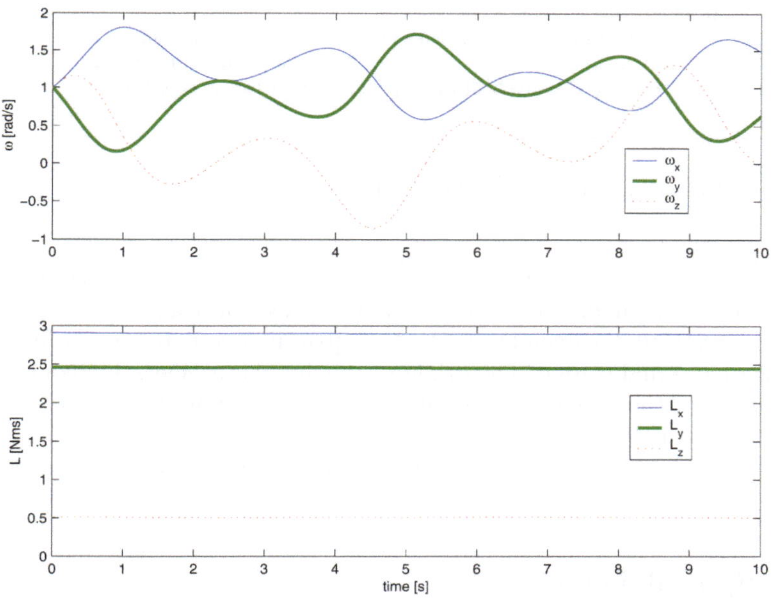

Fig. 6.3 Angular velocity and angular momentum of free rotating rigid body: The angular velocity changes while the angular momentum remains constant

Suppose a cuboid with dimensions $0.1 \times 0.4 \times 0.9$ [m^3] is freely rotating in zero-G space around its center of mass. The angular velocity $\boldsymbol{\omega}$ changes based on Euler's equation (6.4), whereas its posture $\boldsymbol{R}$ changes by the equation of rotation (2.33) from Chapter 2

$$\dot{\boldsymbol{\omega}} = -\boldsymbol{I}^{-1}(\boldsymbol{\omega} \times \boldsymbol{I}\boldsymbol{\omega}) \tag{6.5}$$

$$\dot{\boldsymbol{R}} = \widehat{\boldsymbol{\omega}}\boldsymbol{R}. \tag{6.6}$$

We can integrate these differential equations numerically, assuming the angular acceleration and velocity $\dot{\boldsymbol{\omega}}$, $\boldsymbol{\omega}$ are constant during the short period of Δt using

$$\begin{cases} \boldsymbol{\omega}(t+\Delta t) = \boldsymbol{\omega}(t) + \dot{\boldsymbol{\omega}}\Delta t \\ \boldsymbol{R}(t+\Delta t) = e^{\widehat{\omega}\Delta t}\boldsymbol{R}(t). \end{cases}$$

The simulated result is illustrated in Fig. 6.2. While no external forces exist, we see the object rotates in a complicated manner. The upper graph of Fig. 6.3 shows the change of the angular velocity start from its initial value $\boldsymbol{\omega} = [1\,1\,1]^T$(rad/s). On the other hand, the angular momentum shown in the lower graph of Fig. 6.3 keeps constant. This shows conservation of the angular momentum which is expected, as well as the validity of the simulation.

```
function L = EulerDynamics(j)
global uLINK
I = uLINK(j).R * uLINK(j).I * uLINK(j).R';      % Inertia tensor
L = I * uLINK(j).w;                              % Angular momentum
uLINK(j).dw  = I  (-cross(uLINK(j).w, L));     % Euler's equation
```

Fig. 6.4 EulerDynamics.m Calculation of Euler's equation

Like this simulation, generally, a floating rigid body in 3D space rotates in a complicated manner which does not have a fixed rotating axis[1]. The change of angular velocity is given by (6.5) which gives non-zero angular acceleration with zero external torque. The origin of the angular acceleration comes from the change of the inertia tensor, that is, the relocation of mass distribution caused by rotation.

Figures 6.4 and 6.5 show the Matlab code for the simulation of a rotating object.

6.2 Spatial Velocity

6.2.1 Speed of Rigid Body

In robotics, we use two different ways to represent the translational speed of a rigid body in 3D space.

(A) Speed of reference point: Set an appropriate reference point on a rigid body (ex. the center of mass or the center of the joint), then its speed in the world frame $\boldsymbol{v}_1$ is regarded as the speed of the rigid body. For the

[1] We might have to take care of such rotation to clean up *space debris* orbiting the earth which potentially could harm future space crafts.

```
global uLINK
lx  = 0.1; ly  = 0.4; lz  = 0.9;           % Depth, Width, Height [m]
mass = 36.0;                               % Mass [kg]
MakeRigidBody(1, [lx ly lz], mass);        % Create rigid body

uLINK.p = [0.0, 0.0, 0]'; % Initial position [m]
uLINK.R = eye(3);         % Initial posture
uLINK.w = [1, 1, 1]';     % Initial angular velocity [rad/s]
Dtime   = 0.02;           % Integration time [s]
EndTime = 5.0;            % End of simulation [s]
time   = 0:Dtime:EndTime;
figure
AX=[-0.5 0.5]; AY=[-0.5 0.5]; AZ=[-0.5 1.0]; % 3D view area
for n = 1:length(time)
    L = EulerDynamics(1);       % Euler's equation
    uLINK(1).R = Rodrigues(uLINK(1).w, Dtime) * uLINK(1).R; % Rodrigues
    uLINK(1).w = uLINK(1).w + Dtime * uLINK(1).dw;          % Euler-method
    ShowObject;                       %  Show object
end
```

Fig. 6.5 Simulation code for a free rigid body rotation. See section 6.6.2 for the subroutines MakeRigidBody and ShowObject.

position of reference point $\boldsymbol{p}_1$, the speed of an arbitrary point $\boldsymbol{p}$ on the rigid body is given by

$$\boldsymbol{v}(\boldsymbol{p}) = \boldsymbol{v}_1 + \boldsymbol{\omega} \times (\boldsymbol{p} - \boldsymbol{p}_1).$$

(B) Spatial velocity: The following vector is regarded as the translational speed of a rigid body [30]

$$\boldsymbol{v}_o = \boldsymbol{v}_1 - \boldsymbol{\omega} \times \boldsymbol{p}_1. \tag{6.7}$$

With given instantaneous state of a rigid body, we obtain a unique $\boldsymbol{v}_o$ despite of the choice of $\boldsymbol{p}_1$ as illustrated in Fig. 6.6. Therefore, we can regard $\boldsymbol{v}_o$ as an intrinsic translational speed of a rigid body. In a world frame, the speed of a point $\boldsymbol{p}$ on a rigid body moving at $(\boldsymbol{v}_o, \boldsymbol{\omega})$ is given by

$$\boldsymbol{v}(\boldsymbol{p}) = \boldsymbol{v}_o + \boldsymbol{\omega} \times \boldsymbol{p}. \tag{6.8}$$

A six dimensional vector made of $(\boldsymbol{v}_o, \boldsymbol{\omega})$ is called a *spatial velocity* of a rigid body[30, 101].

Usually, translational speed of a rigid body is represented by the way of (A). Indeed, in chapter 2, we have defined translational speed by (A) taking a local frame origin as a reference point.

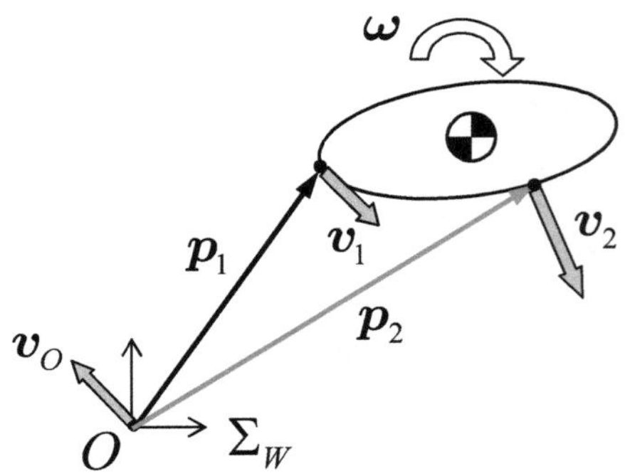

Fig. 6.6 Definition of spatial velocity: $\boldsymbol{v}_o = \boldsymbol{v}_1 - \boldsymbol{\omega} \times \boldsymbol{p}_1$. If we choose a difference point $\boldsymbol{p}_2$ on the rigid body, $\boldsymbol{v}_o = \boldsymbol{v}_2 - \boldsymbol{\omega} \times \boldsymbol{p}_2$ gives the same value. In a two dimension, when we glue a big grass plate on the rigid body, $\boldsymbol{v}_o$ is the speed of grass passing on the origin.

On the other hand, by using the spatial velocity of (B), the treatment of acceleration is much simplified, and it finally results Featherstone's fast dynamic calculation which will be discussed at the end of this section. Therefore, we use the way described in (B) to represent rigid body motions.

6.2.2 Integration of Spatial Velocity

In this section, we explain a method to update the position and the orientation by integrating a given spatial velocity. Unfortunately, this calculation is a little bit complicated. Let us rewrite (6.8) which gives the speed of point $\boldsymbol{p}$ on a rigid body moving at the spatial velocity of $(\boldsymbol{v}_o, \boldsymbol{\omega})$ as

$$\begin{bmatrix} \dot{\boldsymbol{p}} \\ 0 \end{bmatrix} = \begin{bmatrix} \widehat{\boldsymbol{\omega}} & \boldsymbol{v}_o \\ 0\,0\,0 & 0 \end{bmatrix} \begin{bmatrix} \boldsymbol{p} \\ 1 \end{bmatrix} = \boldsymbol{\Xi} \begin{bmatrix} \boldsymbol{p} \\ 1 \end{bmatrix} \tag{6.9}$$

where $\boldsymbol{\Xi}$ is a matrix of 4×4 defined as

$$\boldsymbol{\Xi} \equiv \begin{bmatrix} \widehat{\boldsymbol{\omega}} & \boldsymbol{v}_o \\ 0\,0\,0 & 0 \end{bmatrix}.$$

If $\boldsymbol{\Xi}$ is constant, the solution of the differential equation (6.9) is given by

$$\begin{bmatrix} \boldsymbol{p}(t) \\ 1 \end{bmatrix} = e^{\Xi t} \begin{bmatrix} \boldsymbol{p}(0) \\ 1 \end{bmatrix}. \tag{6.10}$$

The matrix exponential $e^{\Xi t}$ is defined by an infinite series like the rotation matrix shown in Chapter 2

$$e^{\Xi t} = \boldsymbol{E} + \Xi t + \frac{(\Xi t)^2}{2!} + \frac{(\Xi t)^3}{3!} + \dots \quad . \tag{6.11}$$

To simplify this infinite series, we first normalize the spatial velocity so that the norm of the angular velocity becomes one

$$\begin{aligned} \boldsymbol{\omega}_n &= \boldsymbol{\omega}/\|\boldsymbol{\omega}\| \\ \boldsymbol{v}_{on} &= \boldsymbol{v}_o/\|\boldsymbol{\omega}\| \\ t' &= \|\boldsymbol{\omega}\|t. \end{aligned}$$

By using this, we obtain the following equation[2]

$$e^{\Xi t} = \begin{bmatrix} e^{\widehat{\omega}_n t'} & (\boldsymbol{E} - e^{\widehat{\omega}_n t'})(\boldsymbol{\omega}_n \times \boldsymbol{v}_{on}) + \boldsymbol{\omega}_n \boldsymbol{\omega}_n^T \boldsymbol{v}_{on} t' \\ 0\,0\,0 & 1 \end{bmatrix}. \tag{6.12}$$

If the angular velocity $\boldsymbol{\omega}$ is zero, we cannot normalize the spatial velocity. However, such case gives a simple translation without rotation, thus we get

$$e^{\Xi t} = \begin{bmatrix} \boldsymbol{E} & \boldsymbol{v}_o t \\ 0\,0\,0 & 1 \end{bmatrix}. \tag{6.13}$$

When the position and orientation of a rigid body was $(\boldsymbol{p}(t), \boldsymbol{R}(t))$ at time t and its spatial velocity was given as Ξ, its configuration at $t + \Delta t$ is given by

$$\begin{bmatrix} \boldsymbol{R}(t+\Delta t) & \boldsymbol{p}(t+\Delta t) \\ 0\,0\,0 & 1 \end{bmatrix} = e^{\Xi \Delta t} \begin{bmatrix} \boldsymbol{R}(t) & \boldsymbol{p}(t) \\ 0\,0\,0 & 1 \end{bmatrix}. \tag{6.14}$$

Equations (6.12),(6.13) and (6.14) can be coded as a Matlab program shown in Fig. 6.7. In this program a new variable uLINK.vo is introduced to hold the linear part of the spatial velocity $\boldsymbol{v}_o$.

Figure 6.8 shows the motion of a rigid body under a constant spatial velocity calculated by the program of Fig. 6.7.

6.3 Dynamics of Rigid Body

In this section, we discuss the dynamics and the simulation method of a rigid body in 3D space.

6.3.1 Newton-Euler Equations

It is well known that the dynamics of rigid body can be separated into the translation of the center of mass (CoM) and the rotation around it. Thus the rigid body dynamics in 3D space is given by the Newton's equation for

[2] For more detailed derivation, see the textbook of Murray, Li and Sastry [101](pp.39–42).

```
function [p2, R2] = SE3exp(j, dt)
global uLINK
norm_w = norm(uLINK(j).w);
if norm_w < eps
    p2 = uLINK(j).p + dt * uLINK(j).vo;
    R2 = uLINK(j).R;
else
    th = norm_w*dt;
    wn = uLINK(j).w/norm_w;      % normalized vector
    vo = uLINK(j).vo/norm_w;
    rot= Rodrigues(wn, th);
    p2 = rot * uLINK(j).p +(eye(3)-rot)*cross(wn, vo) + wn * wn' * vo * th;
    R2 = rot * uLINK(j).R;
end
```

Fig. 6.7 SE3exp.m Updating position and orientation based on spatial velocity

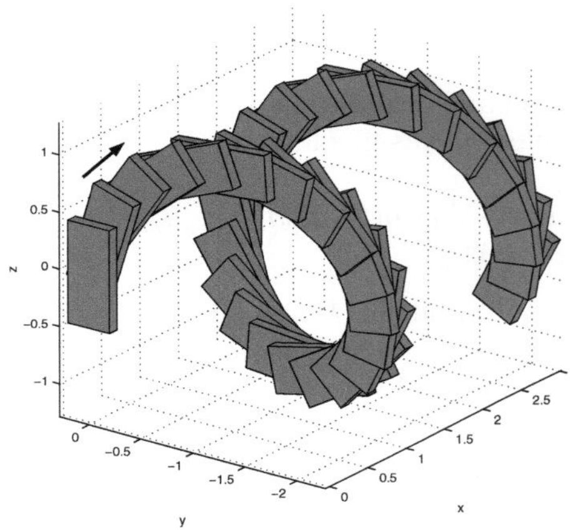

Fig. 6.8 Motion of a rigid body with constant spatial velocity of $\boldsymbol{v}_o = [0.3\,0\,1]^T$(m/s), $\boldsymbol{\omega} = [1\,0\,0]^T$(rad/s). State of every 0.3 second is displayed during the motion of 10 seconds. A trajectory of a rigid body with constant spatial velocity generally becomes a spiral.

the CoM and the Euler's equation around it. These equations are called *Newton-Euler equations* and are given by

$$\boldsymbol{f} = m\ddot{\boldsymbol{c}} \tag{6.15}$$

$$\boldsymbol{\tau}^{(c)} = \boldsymbol{I}\dot{\boldsymbol{\omega}} + \boldsymbol{\omega} \times \boldsymbol{I}\boldsymbol{\omega}, \tag{6.16}$$

where $\boldsymbol{f}$ is the external linear force acting on the CoM, m is the mass of the rigid body, $\boldsymbol{c}$ is the position of CoM with respect to the world frame. $\boldsymbol{\tau}^{(c)}$ is the external moment around CoM, and $\boldsymbol{I}, \boldsymbol{\omega}$ are the inertia tensor and angular velocity represented in the world frame.

6.3.2 Dynamics by Spatial Velocity

Let us rewrite the Newton-Euler equations with spatial velocity. Suppose we have a rigid body with CoM speed of $\boldsymbol{c}$ and rotating at $\boldsymbol{\omega}$. Its spatial velocity is calculated by (6.7) as

$$\boldsymbol{v}_o = \dot{\boldsymbol{c}} - \boldsymbol{\omega} \times \boldsymbol{c}. \tag{6.17}$$

By differentiating this, we obtain

$$\dot{\boldsymbol{v}}_o = \ddot{\boldsymbol{c}} - \dot{\boldsymbol{\omega}} \times \boldsymbol{c} - \boldsymbol{\omega} \times \dot{\boldsymbol{c}}. \tag{6.18}$$

From (6.18) and (6.17), the acceleration of CoM is given as

$$\ddot{\boldsymbol{c}} = \dot{\boldsymbol{v}}_o - \boldsymbol{c} \times \dot{\boldsymbol{\omega}} + \boldsymbol{\omega} \times (\boldsymbol{v}_o + \boldsymbol{\omega} \times \boldsymbol{c}). \tag{6.19}$$

By substituting this into Newton's equation (6.15), we obtain a translational motion equation under the spatial velocity representation

$$\boldsymbol{f} = m(\dot{\boldsymbol{v}}_o - \boldsymbol{c} \times \dot{\boldsymbol{\omega}} + \boldsymbol{\omega} \times (\boldsymbol{v}_o + \boldsymbol{\omega} \times \boldsymbol{c})). \tag{6.20}$$

On the other hand, the force $\boldsymbol{f}$ acting on the CoM and the moment $\boldsymbol{\tau}^{(c)}$ around the CoM create the moment around the origin of the world frame as

$$\boldsymbol{\tau} = \boldsymbol{\tau}^{(c)} + \boldsymbol{c} \times \boldsymbol{f}. \tag{6.21}$$

By substituting Euler's equation (6.16) and (6.20) into (6.21), we obtain a rotational dynamics under the spatial velocity representation

$$\boldsymbol{\tau} = \boldsymbol{I}\dot{\boldsymbol{\omega}} + \boldsymbol{\omega} \times \boldsymbol{I}\boldsymbol{\omega} + m\boldsymbol{c} \times (\dot{\boldsymbol{v}}_o - \boldsymbol{c} \times \dot{\boldsymbol{\omega}} + \boldsymbol{\omega} \times (\boldsymbol{v}_o + \boldsymbol{\omega} \times \boldsymbol{c})). \tag{6.22}$$

The nasty looking equations (6.20) and (6.22) can be organized as the following matrix equation. This is the equation of rigid body motion using spatial velocity

$$\begin{bmatrix} \boldsymbol{f} \\ \boldsymbol{\tau} \end{bmatrix} = \boldsymbol{I}^S \begin{bmatrix} \dot{\boldsymbol{v}}_o \\ \dot{\boldsymbol{\omega}} \end{bmatrix} + \begin{bmatrix} \widehat{\boldsymbol{\omega}} & \boldsymbol{0} \\ \widehat{\boldsymbol{v}}_o & \widehat{\boldsymbol{\omega}} \end{bmatrix} \boldsymbol{I}^S \begin{bmatrix} \boldsymbol{v}_o \\ \boldsymbol{\omega} \end{bmatrix}. \tag{6.23}$$

The matrix $\boldsymbol{I}^S$ is a 6×6 symmetric matrix defined by the following equation and is called spatial inertia matrix

$$\boldsymbol{I}^S \equiv \begin{bmatrix} m\boldsymbol{E} & m\widehat{\boldsymbol{c}}^T \\ m\widehat{\boldsymbol{c}} & m\widehat{\boldsymbol{c}}\widehat{\boldsymbol{c}}^T + \boldsymbol{I} \end{bmatrix}. \tag{6.24}$$

Also, the six dimensional vector $[\dot{\boldsymbol{v}}_o^T, \dot{\boldsymbol{\omega}}^T]^T$ is called the spatial acceleration.

Note that, the momentum and the angular momentum of a rigid body is calculated by

$$\begin{bmatrix} \mathcal{P} \\ \mathcal{L} \end{bmatrix} = \boldsymbol{I}^S \begin{bmatrix} \boldsymbol{v}_o \\ \boldsymbol{\omega} \end{bmatrix}. \tag{6.25}$$

6.3.3 Rigid Body Simulation Based on Spatial Velocity

By using the equation of rigid body motion using spatial velocity (6.23), we can calculate the spatial acceleration under the given force and moment $(\boldsymbol{f}, \boldsymbol{\tau})$ using

$$\begin{bmatrix} \dot{\boldsymbol{v}}_o \\ \dot{\boldsymbol{\omega}} \end{bmatrix} = (\boldsymbol{I}^S)^{-1} \left(\begin{bmatrix} \boldsymbol{f} \\ \boldsymbol{\tau} \end{bmatrix} - \begin{bmatrix} \widehat{\boldsymbol{\omega}} & \boldsymbol{0} \\ \widehat{\boldsymbol{v}}_o & \widehat{\boldsymbol{\omega}} \end{bmatrix} \boldsymbol{I}^S \begin{bmatrix} \boldsymbol{v}_o \\ \boldsymbol{\omega} \end{bmatrix} \right). \tag{6.26}$$

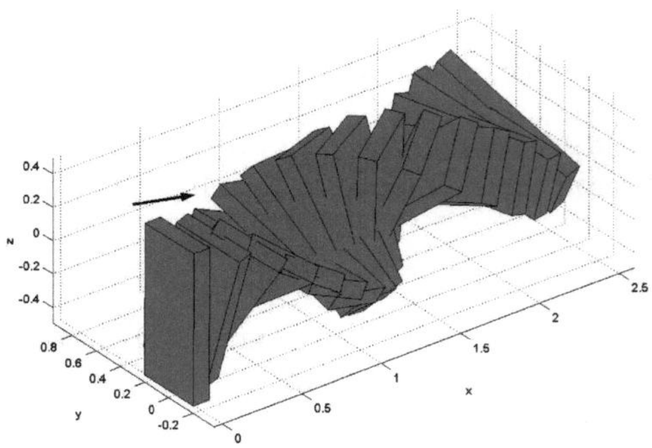

Fig. 6.9 Rigid body dynamics simulation in zero-G space. The motion of five seconds under the initial condition of $\boldsymbol{v}_o = [0.5\, 0.1\, 0]^T$(m/s), $\boldsymbol{\omega} = [1\, 0\, 1]^T$(rad/s) and zero external force/torque (configurations of every 0.3 second are shown).

Figure 6.9 shows the result of simulation of floating rigid body in zero-G space based on this equation ($\boldsymbol{f} = \boldsymbol{\tau} = 0$). The center of mass of the object moves at constant speed with free rotation observed in section 6.1.2.

The spatial velocity $\boldsymbol{v}_o$ and the linear momentum of the simulation is shown in Fig. 6.10. Although the speed of CoM is constant, we have time varying spatial velocity $\boldsymbol{v}_o$ as the result of the object rotation. Of course the linear momentum is kept constant.

Rather than using spatial velocity and (6.26), some people may prefer to use the original Newton-Euler equations (6.15) and (6.16) for such single

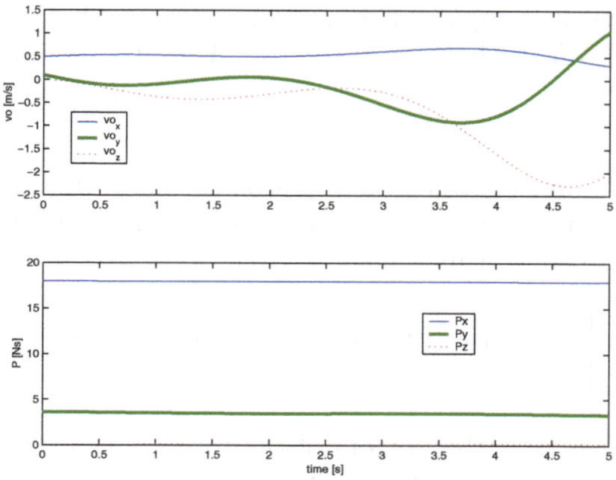

Fig. 6.10 The spatial velocity $\boldsymbol{v}_o$ and the linear momentum $\mathcal{P}$ of the simulation of Fig. 6.9. $\boldsymbol{v}_o$ changes in time while $\mathcal{P}$ keeps constant corresponding the linear constant speed of the center of mass.

```
function [P,L] = SE3dynamics(j)
global uLINK
w_c = uLINK(j).R * uLINK(j).c + uLINK(j).p;     % Center of mass
w_I = uLINK(j).R * uLINK(j).I * uLINK(j).R';    % Inertia tensor
c_hat = wedge(w_c);
Iww = w_I + uLINK(j).m * c_hat * c_hat';
Ivv = uLINK(j).m * eye(3);
Iwv = uLINK(j).m * c_hat;
P = uLINK(j).m * (uLINK(j).vo + cross(uLINK(j).w,w_c));       % Lin. momentum
L = uLINK(j).m * cross(w_c,uLINK(j).vo) + Iww * uLINK(j).w; % Ang. momentum
pp = [cross(uLINK(j).w,P);
      cross(uLINK(j).vo,P) + cross(uLINK(j).w,L)];
a0 = -[Ivv, Iwv'; Iwv, Iww]  pp;                         % Spatial acceleration
uLINK(j).dvo = a0(1:3);
uLINK(j).dw  = a0(4:6);
```

Fig. 6.11 SE3dynamics.m Equation of rigid body motion

rigid body simulation. Nevertheless please be patient until Section 6.4 where the advantage of the spatial velocity representation will be shown by the calculation of multi-body system.

A part of Matlab code used for this simulation is listed in Fig. 6.11. A new variable uLINK.dvo is introduced to hold the spatial acceleration $\dot{\boldsymbol{v}}_o$.

6.3.4 Simulation of a Spinning Top

Based on the above discussion, let us simulate a spinning top which performs a very interesting behavior of a rigid body in 3D space.

The forces acts on a top are (1) gravity on the CoM $\boldsymbol{f}_g$ and (2) the floor reaction force act on the bottom of the top. The first force is given as

$$\boldsymbol{f}_g = [0\ 0\ \ -mg]^T.$$

The second force only acts when the top is on the floor. Its vertical component can be calculated by a spring-damper system and the horizontal frictional forces can be calculated using

$$\boldsymbol{f}_c = \begin{cases} [-D_f\dot{p}_x,\ -D_f\dot{p}_y,\ -K_f p_z - D_f\dot{p}_z]^T & (\text{if } p_z < 0) \\ [0,\ 0,\ 0]^T & (\text{if } p_z > 0) \end{cases}$$

where $\boldsymbol{p} \equiv [p_x\ p_y\ p_z]^T$ is the bottom of the top. D_f is a damping coefficient and K_f is a spring coefficient to give the floor stiffness.

These forces generates the moment $\boldsymbol{\tau}$ around the origin of the world frame as

$$\boldsymbol{\tau} = \boldsymbol{c} \times \boldsymbol{f}_g + \boldsymbol{p} \times \boldsymbol{f}_c.$$

See Section 6.6.1 to know more about this calculation.

The process of force generation described above is programmed as Fig. 6.12.

```
function [f,t] = TopForce(j)
global uLINK G Kf Df
w_c = uLINK(j).R * uLINK(j).c + uLINK(j).p;    % center of mass
f = [0 0 -uLINK(j).m * G]';    % gravity
t = cross(w_c, f);             % gravity moment around the origin

if uLINK(j).p(3) < 0.0  % the top is contacting
    v = uLINK(j).vo + cross(uLINK(j).w,uLINK(j).p);  % contacting speed
    fc = [-Df*v(1)  -Df*v(2)  -Kf*uLINK(j).p(3)-Df*v(3)]';
    f  = f + fc;
    t  = t + cross(uLINK(j).p, fc);
end
```

Fig. 6.12 TopForce.m : Force and moment acting on the top

The simulation of a top falling in the gravity of 1G with appropriate initial angular velocity is shown in Fig. 6.13. We can observe the precession which appears as a circular motion of the upper end of the spinning shaft.

The simulation code is shown in Fig. 6.14. The spatial acceleration is calculated by (6.26) and the spatial velocity is updated using

$$\begin{cases} \boldsymbol{v}_o(t+\Delta t) = \boldsymbol{v}_o(t) + \dot{\boldsymbol{v}}_o\Delta t \\ \boldsymbol{\omega}(t+\Delta t) = \boldsymbol{\omega}(t) + \dot{\boldsymbol{\omega}}\Delta t. \end{cases}$$

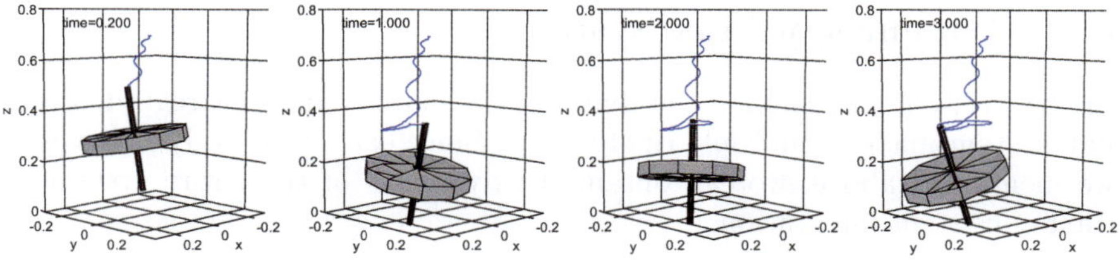

Fig. 6.13 Simulation of a top falling onto a floor with initial angular velocity of $[0\,0\,50]^T$(rad/s). A locus of the upper end of the spinning shaft is shown to visualize the precession.

```
global uLINK G Kf Df
G = 9.8;
Kf = 1.0E+4; Df = 1.0E+3;      % Floor stiffness[N/m] viscosity[N/(m/s)]
r = 0.2; a = 0.05; c = 0.2;    % Top radius, thickness, shaft length/2 [m]
MakeTop(1, r,a,c);

uLINK(1).p = [0 0 0.3]';                  % Initial position [m]
uLINK(1).R = Rodrigues([1 0 0],pi/50);    % Initial posture
uLINK(1).vo= [0 0 0]';                    % Initial speed [m/s]
uLINK(1).w = [0 0 50]';                   % Initial spin [rad/s]
Dtime   = 0.002;
EndTime = 2.0;
time    = 0:Dtime:EndTime;

figure
frame_skip = 3;
AX=[-0.2 0.4];  AY=[-0.3 0.3]; AZ=[0 0.8];  % 3D view area
for n = 1:length(time)
    [f,tau] = TopForce(1);                              % External force
    [P,L]   = SE3dynamics(1,f,tau);                     % Acceleration
    [uLINK.p, uLINK.R] = SE3exp(1, Dtime);              % Update configuration
    uLINK(1).w = uLINK(1).w + Dtime * uLINK(1).dw;     % Update ang. velocity
    uLINK(1).vo= uLINK(1).vo+ Dtime * uLINK(1).dvo;    % Update lin. velocity
    if mod(n,frame_skip) == 0
        ShowObject;                                     % Show the top
    end
end
```

Fig. 6.14 Matlab code to simulate a spinning top. It also shows an animation of the simulated top. See section 6.6.2 for subroutines `MakeTop` and `ShowObject`.

The position and orientation of the top is updated by using the program of Fig. 6.7 based on (6.14). The calculated top motion is simultaneously displayed.

6.4 Dynamics of Link System

In this section, we examine the dynamics of humanoid robot regarded as a link system made of multiple rigid bodies. Since each rigid body behaves as we mentioned above, we can obtain the dynamics of the entire system by their consistent integration.

6.4.1 Forward Kinematics with Acceleration

First, we examine a pair of rigid bodies connected by a joint, which represents a minimum element of a humanoid robot.

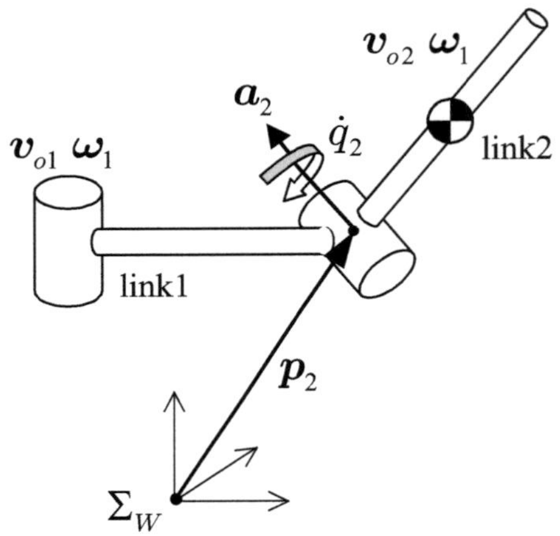

Fig. 6.15 Propagation of spatial velocity: link2 obtains relative speed with respect to link1 by rotating the joint axis $\boldsymbol{a}_2$ at the speed of $\dot{q}_2$

Suppose we have the link1 in the space connected with link2 via rotational joint as illustrated in Fig. 6.15. When the link1 is stationary, by rotating the joint $\boldsymbol{a}_2$ at the speed of $\dot{q}_2$, link2 gets angular velocity of

$$\boldsymbol{\omega}_2 = \boldsymbol{a}_2\dot{q}_2. \tag{6.27}$$

In the spatial velocity representation of Section 6.2, the linear velocity of link2 is

$$\boldsymbol{v}_{o2} = \dot{\boldsymbol{p}}_2 - \boldsymbol{\omega}_2 \times \boldsymbol{p}_2$$

where $\boldsymbol{p}_2$ is the origin of the link2.

Since we are assuming that link1 is stationary, $\dot{\boldsymbol{p}}_2 = 0$. By substituting (6.27) into the above equation, we get

$$\boldsymbol{v}_{o2} = \boldsymbol{p}_2 \times \boldsymbol{a}_2\dot{q}_2. \tag{6.28}$$

Equations (6.28) and (6.27) give the relative speed $(\boldsymbol{v}_{o2}, \boldsymbol{\omega}_2)$ of link2 with respect to link1.

When the link1 have the spatial velocity $(\boldsymbol{v}_{o1}, \boldsymbol{\omega}_1)$, we can simply add it and the spatial velocity of link2 becomes

$$\begin{bmatrix} \boldsymbol{v}_{o2} \\ \boldsymbol{\omega}_2 \end{bmatrix} = \begin{bmatrix} \boldsymbol{v}_{o1} \\ \boldsymbol{\omega}_1 \end{bmatrix} + \begin{bmatrix} \boldsymbol{p}_2 \times \boldsymbol{a}_2 \\ \boldsymbol{a}_2 \end{bmatrix} \dot{q}_2. \tag{6.29}$$

Hereafter, we use following notation to simplify our equations

$$\boldsymbol{\xi}_2 = \boldsymbol{\xi}_1 + \boldsymbol{s}_2 \dot{q}_2 \tag{6.30}$$

$$\boldsymbol{\xi}_j \equiv \begin{bmatrix} \boldsymbol{v}_{oj} \\ \boldsymbol{\omega}_j \end{bmatrix}, \quad \boldsymbol{s}_j \equiv \begin{bmatrix} \boldsymbol{p}_j \times \boldsymbol{a}_j \\ \boldsymbol{a}_j \end{bmatrix}$$

where $\boldsymbol{\xi}_j$ is the spatial velocity and $\boldsymbol{s}_j$ is the spatial velocity generated by the joint rotation of unit speed. Both of them are vectors of six dimension.

By differentiating (6.30), we obtain the propagation rule of spatial acceleration[3]

$$\dot{\boldsymbol{\xi}}_2 = \dot{\boldsymbol{\xi}}_1 + \dot{\boldsymbol{s}}_2 \dot{q}_2 + \boldsymbol{s}_2 \ddot{q}_2 \tag{6.31}$$

where $\dot{\boldsymbol{s}}_2$ can be calculated using

$$\dot{\boldsymbol{s}}_2 = \begin{bmatrix} \widehat{\boldsymbol{\omega}}_1 & \widehat{\boldsymbol{v}}_{o1} \\ \boldsymbol{0} & \widehat{\boldsymbol{\omega}}_1 \end{bmatrix} \boldsymbol{s}_2. \tag{6.32}$$

Let us assume that we know the position and orientation, the spatial velocity and the spatial acceleration of the robot body. Then if the angles, the velocity and the acceleration of all joints are also known, we can calculate the spatial velocities and the spatial accelerations of all links by applying (6.30) and (6.31) from the body toward the end of limbs. This calculation can be easily programmed by using recursive algorithm as Fig. 6.16.

6.4.2 Inverse Dynamics of Link System

Next we consider the force and the moment act on each link. For preparation, let us rewrite the equation of rigid body motion we obtained in section 6.3 by using the spatial velocity vector $\boldsymbol{\xi}$

$$\begin{bmatrix} \boldsymbol{f} \\ \boldsymbol{\tau} \end{bmatrix} = \boldsymbol{I}^S \dot{\boldsymbol{\xi}} + \boldsymbol{\xi} \times \boldsymbol{I}^S \boldsymbol{\xi} \tag{6.33}$$

where the operator $\times$ represents the following calculation.

[3] Thanks to the spatial velocity representation we get this simple equation. When we use a conventional speed of a reference point, we need a more complicated propagation rule of acceleration[99].

```
function ForwardAllKinematics(j)
global uLINK G
if j == 0 return; end
if j ~= 1
    mom = uLINK(j).mother;
    %% Position and orientation
    uLINK(j).p = uLINK(mom).R * uLINK(j).b + uLINK(mom).p;
    uLINK(j).R = uLINK(mom).R * Rodrigues(uLINK(j).a, uLINK(j).q);
    %% Spatial velocity
    sw = uLINK(mom).R * uLINK(j).a;  % axis vector in world frame
    sv = cross(uLINK(j).p, sw);      % p_i x axis
    uLINK(j).w  = uLINK(mom).w  + sw * uLINK(j).dq;
    uLINK(j).vo = uLINK(mom).vo + sv * uLINK(j).dq;
    %% Spatial acceleration
    dsv = cross(uLINK(mom).w, sv) + cross(uLINK(mom).vo, sw);
    dsw = cross(uLINK(mom).w, sw);
    uLINK(j).dw  = uLINK(mom).dw  + dsw * uLINK(j).dq + sw * uLINK(j).ddq;
    uLINK(j).dvo = uLINK(mom).dvo + dsv * uLINK(j).dq + sv * uLINK(j).ddq;
    uLINK(j).sw = sw;                % store h1 and h2 for future use
    uLINK(j).sv = sv;
end
ForwardAllKinematics(uLINK(j).sister); % Propagate to sister
ForwardAllKinematics(uLINK(j).child);  % Propagate to child
```

Fig. 6.16 Calculate spatial velocity and acceleration of all links

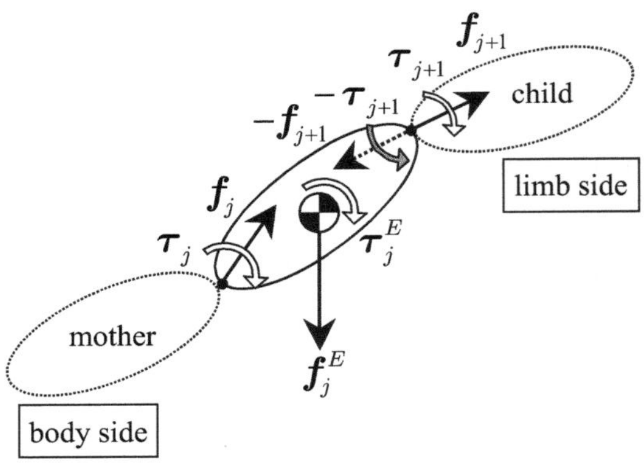

Fig. 6.17 Force and moment acting on the j-th link

$$\boldsymbol{\xi}\times = \begin{bmatrix} \boldsymbol{v}_o \\ \boldsymbol{\omega} \end{bmatrix} \times \equiv \begin{bmatrix} \widehat{\boldsymbol{\omega}} & \mathbf{0} \\ \widehat{\boldsymbol{v}}_o & \widehat{\boldsymbol{\omega}} \end{bmatrix}. \tag{6.34}$$

Figure 6.17 shows the force and the moment acting on link j. $\boldsymbol{f}_j, \boldsymbol{\tau}_j$ are the force and the moment from body side (mother) to link i. The environmental force and moment directory acting on link i are $\boldsymbol{f}_j^E, \boldsymbol{\tau}_j^E$. Effects from the mother link, environment and reaction effect from the child link cause the motion of the link i. The equation of motion becomes

$$\begin{bmatrix} \boldsymbol{f}_j \\ \boldsymbol{\tau}_j \end{bmatrix} + \begin{bmatrix} \boldsymbol{f}_j^E \\ \boldsymbol{\tau}_j^E \end{bmatrix} - \begin{bmatrix} \boldsymbol{f}_{j+1} \\ \boldsymbol{\tau}_{j+1} \end{bmatrix} = \boldsymbol{I}_j^S \dot{\boldsymbol{\xi}}_j + \boldsymbol{\xi}_j \times \boldsymbol{I}_j^S \boldsymbol{\xi}_j. \tag{6.35}$$

By rewriting this, we obtain a recurrence equation to give the propagation of the force and moment

$$\begin{bmatrix} \boldsymbol{f}_j \\ \boldsymbol{\tau}_j \end{bmatrix} = \boldsymbol{I}_j^S \dot{\boldsymbol{\xi}}_j + \boldsymbol{\xi}_j \times \boldsymbol{I}_j^S \boldsymbol{\xi}_j - \begin{bmatrix} \boldsymbol{f}_j^E \\ \boldsymbol{\tau}_j^E \end{bmatrix} + \begin{bmatrix} \boldsymbol{f}_{j+1} \\ \boldsymbol{\tau}_{j+1} \end{bmatrix}. \tag{6.36}$$

```
function [f,t] = InverseDynamics(j)
global uLINK
if j == 0
    f=[0,0,0]';
    t=[0,0,0]';
    return;
end
c = uLINK(j).R * uLINK(j).c + uLINK(j).p;     % Center of mass
I = uLINK(j).R * uLINK(j).I * uLINK(j).R';    % Inertia tensor
c_hat = hat(c);
I = I + uLINK(j).m * c_hat * c_hat';
P = uLINK(j).m * (uLINK(j).vo + cross(uLINK(j).w,c));    % Momentum
L = uLINK(j).m * cross(c,uLINK(j).vo) + I * uLINK(j).w; % Ang.momentum
f0 = uLINK(j).m * (uLINK(j).dvo + cross(uLINK(j).dw,c)) ...
     + cross(uLINK(j).w,P);
t0 = uLINK(j).m * cross(c,uLINK(j).dvo) + I * uLINK(j).dw ...
     + cross(uLINK(j).vo,P) + cross(uLINK(j).w,L);

[f1,t1] = InverseDynamics(uLINK(j).child);    % Force and moment form child
f = f0 + f1;
t = t0 + t1;
if j ~= 1
    uLINK(j).u = uLINK(j).sv' * f + uLINK(j).sw' * t;  % Joint torque
end
[f2,t2] = InverseDynamics(uLINK(j).sister);   % Force and moment from sister
f = f + f2;
t = t + t2;
```

Fig. 6.18 Code for inverse dynamics

When the link j was the extreme link, $\boldsymbol{f}_{j+1}, \boldsymbol{\tau}_{j+1}$ becomes zero and we can calculate $\boldsymbol{f}_j, \boldsymbol{\tau}_j$. Therefore, when we start the calculation from the extreme links, we can obtain the force and moment for all links. The torque u_j at each joint axis can be calculated as

$$u_j = \boldsymbol{s}_j^T \begin{bmatrix} \boldsymbol{f}_j \\ \boldsymbol{\tau}_j \end{bmatrix}. \tag{6.37}$$

Those equations are programmed by using recursive algorithm in Fig. 6.18. This program works correctly when the link system has branches.

The calculation shown in this section is called the **inverse dynamics**. By feed-forwarding the joint torque calculated by the inverse dynamics, we can realize a fast and accurate robot motion. Such control technique is called the **computed torque method**. However, to apply the computed torque method, the first link must be fixed on the ground so that the arbitrary force and moment can be generated. This is true in the case of industrial manipulators.

The base link of humanoid robot, the body is not fixed on the ground, thus there is no guarantee that the robot can obtain the force and moment required to keep the desired body posture. If the motion pattern was inappropriate, the position and orientation of the robot body will deviate from the plan. This can be simulated by integrating the body acceleration. It is a special variation of the forward dynamics simulation we will discuss in the next section. In OpenHRP dynamics simulator which was developed in Humanoid Robotics Project (HRP), this is called a simulation of *high-gain mode* [31]. In many cases, such simulation is enough to confirm the stability of the planned walking pattern.

6.4.3 Forward Dynamics of Link System

Figure 6.19 shows the result of simulation in which all joint torques of HRP-2 are set to zero and the robot drops onto the floor.

Fig. 6.19 Simulation of dropping HRP-2 with zero joint torques

Like this simulation, a robot performs various motion by being subjected to force and torque from the environment as well as by its own joint torques. Calculating robot motion under the given joint torques and the external forces is called **forward dynamics**. In this section, we explain the basic calculation of forward dynamics based on the knowledges we have learned in the previous sections.

In general, a humanoid robot with n joints have $n+6$ degree of freedom since the body is regarded as free link in 3D space. Its equation of motion has the following form [140, 77]

$$\boldsymbol{u}_G = \boldsymbol{A}_G \ddot{\boldsymbol{x}}_G + \boldsymbol{b}_G \tag{6.38}$$

where $\boldsymbol{A}_G \in R^{(n+6)\times(n+6)}$ is an inertia matrix, $\boldsymbol{b}_G \in R^{(n+6)}$ is a vector representing the Coriolis and the centrifugal forces and the gravity. $\boldsymbol{u}_G$ is the input of robot, $\ddot{\boldsymbol{x}}_G$ is the acceleration. They are defined as

$$\boldsymbol{u}_G \equiv \begin{bmatrix} \boldsymbol{f}_B \\ \boldsymbol{\tau}_B \\ \boldsymbol{u} \end{bmatrix}, \quad \ddot{\boldsymbol{x}}_G \equiv \begin{bmatrix} \dot{\boldsymbol{v}}_{oB} \\ \dot{\boldsymbol{\omega}}_B \\ \ddot{\boldsymbol{q}} \end{bmatrix}$$

where

$(\boldsymbol{f}_B, \boldsymbol{\tau}_B)$: Force and moment on the body (except gravity)
$(\dot{\boldsymbol{v}}_{oB}, \dot{\boldsymbol{\omega}}_B)$: Spatial acceleration of the body
$\boldsymbol{u}$: All joint torques ($\boldsymbol{u} \in R^n$)
$\ddot{\boldsymbol{q}}$: All joint accelerations ($\ddot{\boldsymbol{q}} \in R^n$).

If we have known $\boldsymbol{A}_G \in R^{(n+6)\times(n+6)}$ and $\boldsymbol{b}_G \in R^{(n+6)}$, the acceleration under the given input $\boldsymbol{u}_G$ can be calculated as

$$\ddot{\boldsymbol{x}}_G = \boldsymbol{A}_G^{-1}(\boldsymbol{u}_G - \boldsymbol{b}_G). \tag{6.39}$$

This is the goal of forward dynamics.

Now, let us define a function $InvDyn()$ which calculates joint torques with given joint accelerations based on the inverse dynamics algorithm explained in the previous section

$$\boldsymbol{u}_G = InvDyn(\ddot{\boldsymbol{x}}_G). \tag{6.40}$$

By comparing (6.38) and (6.40), we see $\boldsymbol{b}_G$ can be calculated from $InvDyn()$ with the acceleration vector $\ddot{\boldsymbol{x}}_G = \boldsymbol{0}$

$$\boldsymbol{b}_G = InvDyn(\boldsymbol{0}). \tag{6.41}$$

Next, we assume a $n+6$ vector $\boldsymbol{\delta}_i$ whose i-th element is one and all other elements are zero. By setting $\ddot{\boldsymbol{x}}_G = \boldsymbol{\delta}_i$, we get

$$\boldsymbol{A}_G \boldsymbol{\delta}_i = InvDyn(\boldsymbol{\delta}_i) - \boldsymbol{b}_G. \tag{6.42}$$

The left hand side of the above equation represents i-th row of the matrix $\boldsymbol{A}_G$. Therefore, if we repeat the calculation with i changing from 1 to $n+6$, all component of $\boldsymbol{A}_G$ can be obtained. This method is called the *unit vector method* and was proposed by Walker and Orin [92].

Once $\boldsymbol{A}_G$ and $\boldsymbol{b}_G$ were obtained, it is straightforward to calculate the acceleration using (6.39). By integrating this, we can perform a dynamic simulation of robot.

Figure 6.20 gives a Matlab code which calculate the spatial acceleration of the body and the joint accelerations based on the unit vector method. The program to calculate *InvDyn*() is given at Fig. 6.29 in the appendix at the end of this chapter.

```
nDoF = length(uLINK)-1+6;
A    = zeros(nDoF,nDoF);
b    = InvDyn(0);
for n=1:nDoF
    A(:,n) = InvDyn(n) - b;
end
% add motor inertia
for n=7:nDoF
    j = n-6+1;
    A(n,n) = A(n,n) + uLINK(j).Ir * uLINK(j).gr^2;
end
u    = [0 0 0 0 0 0 u_joint(2:end)']';
ddq = A   (-b + u);
uLINK(1).dvo = ddq(1:3);
uLINK(1).dw  = ddq(4:6);
for j=1:length(uLINK)-1
    uLINK(j+1).ddq = ddq(j+6);
end
```

Fig. 6.20 Forward dynamics based on the unit vector method

Although the unit vector method is easy to understand and program, its performance quickly decreases when the number of robot joint increases. The first problem is the calculation method of the inertia matrix. For a robot with n joints, the unit vector method performs $n + 7$ calculations of the inverse dynamics to obtain $\boldsymbol{A}_G, \boldsymbol{b}_G$. Since the time for one inverse dynamics calculation is proportional to n, the total calculation time increases in proportional to $n(n + 7)$. This problem can be solved by introducing an algorithm which directly calculates the inertia matrix [92, 30].

The second, and more intrinsic problem is to solve (6.39), a huge simultaneous linear equations. Although the efficient algorithms like Gauss-Jordan elimination or LU decomposition are known [136], it always takes time proportional to $(n + 6)^3$ and it becomes a serious bottle-neck for faster forward dynamics calculation. In the next section, we introduce the method to solve this problem.

6.4.4 Featherstone's Method

Suppose we have two links connected by a revolutional joint floating in 3D space as shown in Fig. 6.21. Assuming that we have already known the spatial acceleration of link1, we want to derive the joint acceleration $\ddot{q}_2$ for a given torque u_2. The spatial acceleration of the link2 is given by

$$\dot{\boldsymbol{\xi}}_2 = \dot{\boldsymbol{\xi}}_1 + \dot{\boldsymbol{s}}_2\dot{q}_2 + \boldsymbol{s}_2\ddot{q}_2. \tag{6.43}$$

From the acceleration, we can calculate the necessary force and moment by using the equation of motion.

$$\begin{bmatrix} \boldsymbol{f}_2 \\ \boldsymbol{\tau}_2 \end{bmatrix} = \boldsymbol{I}_2^S\dot{\boldsymbol{\xi}}_2 + \boldsymbol{\xi}_2 \times \boldsymbol{I}_2^S\boldsymbol{\xi}_2 \tag{6.44}$$

Moreover, the joint torque u_2 is a projection of this force and moment

$$u_2 = \boldsymbol{s}_2^T \begin{bmatrix} \boldsymbol{f}_2 \\ \boldsymbol{\tau}_2 \end{bmatrix}. \tag{6.45}$$

Substituting (6.43) and (6.44) into this equation and we can solve it for the joint acceleration

$$\ddot{q}_2 = \frac{u_2 - \boldsymbol{s}_2^T(\boldsymbol{I}_2^S(\dot{\boldsymbol{\xi}}_1 + \dot{\boldsymbol{s}}_2\dot{q}_2) + \boldsymbol{\xi}_2 \times \boldsymbol{I}_2^S\boldsymbol{\xi}_2)}{\boldsymbol{s}_2^T\boldsymbol{I}_2^S\boldsymbol{s}_2}. \tag{6.46}$$

By this equation, we can immediately calculate the joint acceleration $\ddot{q}_2$ for the given joint torque u_2 without dealing with the inertia matrix that appeared in (6.39) of the previous section.

However, we can use (6.46) only if link2 is the outermost link. If we have extra links starting from link2 as in Fig. 6.22, (6.44) will no longer be valid.

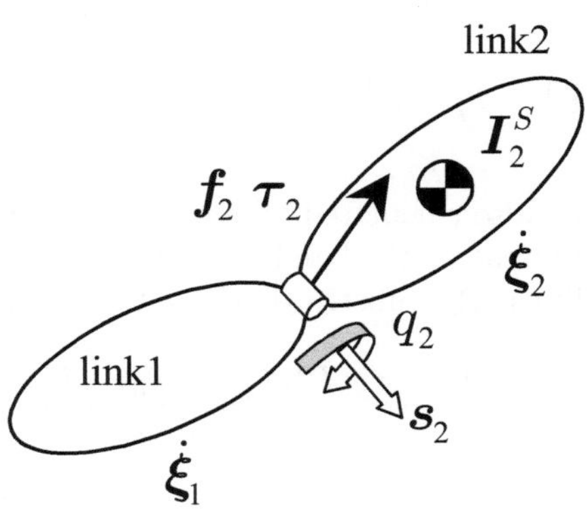

Fig. 6.21 Two links in space

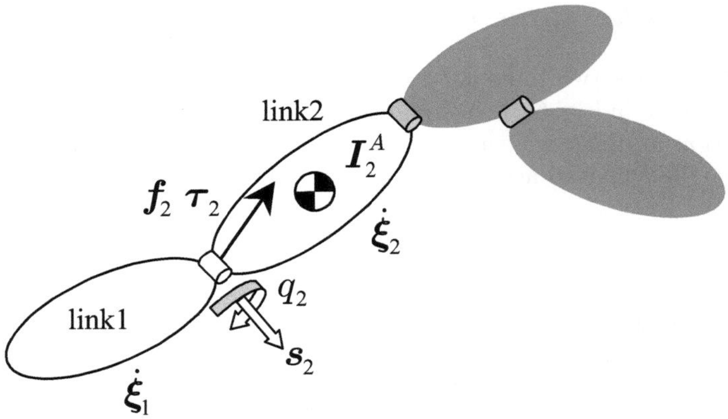

Fig. 6.22 Articulated-Body Inertia: I_2^A represents the relationship between the acceleration and the applied force on the link2. The dynamic effects of all mechanisms (gray part) farther than link2 from the body are taken into consideration.

One way to use (6.46) would be to introduce a new inertia tensor I_2^A which holds all effects of the extra links farther than link2, which satisfies

$$\begin{bmatrix} f_2 \\ \tau_2 \end{bmatrix} = I_2^A \dot{\xi}_2 + b_2. \tag{6.47}$$

I_2^A is called **articulated-body inertia**. An articulated-body inertia matrix represents the relationship between the acceleration and the force acting on the link of interest. b_2 is the bias force, which includes the Coriolis, centrifugal and gravity forces and the joint torques acting on the mechanism farther than the link2.

Featherstone showed that an articulated-body inertia can be calculated by the recursive equation

$$I_j^A = I_j^S + I_{j+1}^A - \frac{I_{j+1}^A s_{j+1} s_{j+1}^T I_{j+1}^A}{s_{j+1}^T I_{j+1}^A s_{j+1}}. \tag{6.48}$$

The bias force can be also calculated by a similar equation.

Suppose we have already calculated the articulated-body inertia and the bias forces for all links. By substituting (6.47) and (6.43) into (6.45), we obtain

$$\ddot{q}_2 = \frac{u_2 - s_2^T (I_2^A (\dot{\xi}_1 + \dot{s}_2 \dot{q}_2) + b_2)}{s_2^T I_2^A s_2}. \tag{6.49}$$

That is, the equation which directly calculates the joint acceleration with the given joint torque. This is the heart of Featherstone's theory.

In summary, the calculation of forward dynamics can be performed by the three steps as following.

1. Calculate the position, orientation and spatial velocity of all links from the body link to the outer links (Forward kinematics).
2. Calculate the articulated-body inertia and the bias force for all links from the outermost links to the body link.
3. Calculate the joint acceleration by (6.49) and the link spatial acceleration by (6.43) from the body link to the outer links.

Since each step takes time in proportion to the number of joints, n, the total calculation time is also proportional to n. Such an algorithm is called to have the speed of order n, or denoted as $O(n)$. For example, the unit vector method in the previous section is $O(n^3)$ since it takes time in proportional to cube of n. We implemented the Featherstone algorithm on Matlab. For HRP-2 ($n = 30$), it was about twenty times faster than the unit vector method.

6.5 Background Material for This Section

The basic theory of the motion of object was established in 17th to 18th century by Newton, Euler and other scientists. However, it was not the end of research.

Indeed, about twenty years ago, an undergraduate student attempted to simulate a biped robot of twelve degree of freedom, so he started to calculate the equation of motion by hand. It took whole summer and he needed a whole notebook to write down the resulted equations. In those days, there existed commercial simulation software which could handle complicated mechanical systems, however, they were extremely expensive and specialized for computers called engineering workstations whose price was prohibitive. Moreover, even using such software, it took hours just to simulate a simple biped robot walking several steps!

Such situation has drastically changed by the efficient and elegant dynamic calculation we explained in this chapter. These algorithms were developed by researchers of robotics and aeronautical engineering in 1980s and 1990s. Currently, the state-of-the-art technology can realize a dynamic simulation with real-time speed. Moreover, researchers are still working for improvement and they recently proposed algorithms of $O(\log(n))$ based on parallel computing [98, 78]. The application of those fast dynamic simulation is not limited to robotics, but they are also used to analyze satellite motion when the solar cell panels expand or to visualize realistic scenes by computer graphics in the movies and interactive computer games.

6.6 Appendix

6.6.1 *Treatment of Force and Moment*

In this section, all equations of motion are represented in the world frame, thus the force and the moment must be written with respect to the world frame. For example, even for the gravity acting on the CoM of an object, we must consider its moment around the world frame origin. Although this seems complicated, we can correctly calculate the interaction between multi rigid bodies.

The basic rules to convert the force and moment into the world frame are as follows.

Rule 1: A force $\boldsymbol{f}$ acting on a point $\boldsymbol{p}$ is converted to a force $\boldsymbol{f}$ and a torque $\boldsymbol{p} \times \boldsymbol{f}$ with respect to the world frame (Fig. 6.23(a)).
Rule 2: A moment $\boldsymbol{\tau}$ acting on a point $\boldsymbol{p}$ is the same $\boldsymbol{\tau}$ with respect to the world frame (Fig. 6.23(b)).

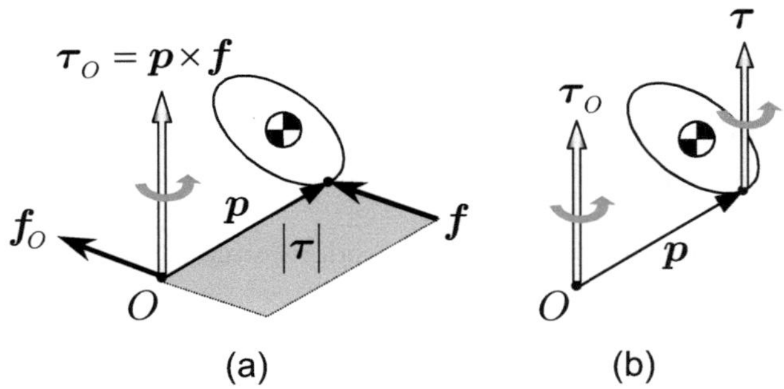

Fig. 6.23 (a)World frame representation of a force acting on a point (b)World frame representation of a moment acting on a point

In general, when a force $\boldsymbol{f}_p$ and a moment $\boldsymbol{\tau}_p$ are simultaneously acting on a point $\boldsymbol{p}$, they are converted a force and moment $\boldsymbol{f}_o, \boldsymbol{\tau}_o$ around the origin given by

$$\begin{bmatrix} \boldsymbol{f}_o \\ \boldsymbol{\tau}_o \end{bmatrix} = \begin{bmatrix} \boldsymbol{f}_p \\ \boldsymbol{p} \times \boldsymbol{f}_p + \boldsymbol{\tau}_p \end{bmatrix}. \tag{6.50}$$

One might think them obvious, we tend to make a curious mistake for this sort of calculation. The quiz shown in Fig. 6.24 might help to understand this problem.

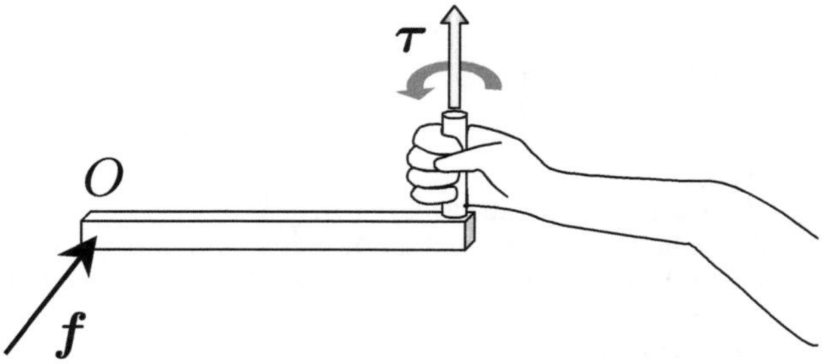

Fig. 6.24 A person is applying a moment $\boldsymbol{\tau}$ gripping the handle of a bar. To prevent the rotation of the bar, it might be enough to apply an appropriate force $\boldsymbol{f}$ at the far side of the bar. Assuming this point to be the origin O, doesn't this violate the rule 2, which asserts the same moment will appear at the origin? (The answer is on page 210).

6.6.2 Subroutines

```
vert = uLINK(1).R * uLINK(1).vertex;           % rotation
for k = 1:3
    vert(k,:) = vert(k,:) + uLINK(1).p(k);    % translation
end
newplot
h = patch('faces',uLINK(1).face','vertices',vert','FaceColor',[0.5 0.5 0.5]);
axis equal; view(3); grid on; xlim(AX); ylim(AY); zlim(AZ);
text(0.25, -0.25, 0.8, ['time=',num2str(time(n),'%5.3f')])
drawnow
```

Fig. 6.25 `ShowObject.m`: Show animation of the object

```
function MakeRigidBody(j, wdh, mass)
global uLINK
uLINK(j).m = mass;                                    % mass
uLINK(j).c = [0 0 0]';                                % center of mass
uLINK(j).I = [ 1/12*(wdh(2)^2 + wdh(3)^2) 0 0;
             0 1/12*(wdh(1)^2 + wdh(3)^2)  0;
             0 0 1/12*(wdh(1)^2 + wdh(2)^2)] * uLINK.m;   % inertia tensor
uLINK(j).vertex = 0.5*[
   -wdh(1) -wdh(2)  -wdh(3);
   -wdh(1)  wdh(2)  -wdh(3);
    wdh(1)  wdh(2)  -wdh(3);
    wdh(1) -wdh(2)  -wdh(3);
   -wdh(1) -wdh(2)   wdh(3);
   -wdh(1)  wdh(2)   wdh(3);
    wdh(1)  wdh(2)   wdh(3);
    wdh(1) -wdh(2)   wdh(3);
]';                                         % vertex points
uLINK(1).face = [
   1 2 3 4; 2 6 7 3; 4 3 7 8; 1 5 8 4; 1 2 6 5; 5 6 7 8;
]';                                         % polygons
```

Fig. 6.26 `MakeRigidBody.m`: Setup a monolith and its parameters

```
function MakeTop(j, r,a,c)
global uLINK

[vertex1,face1] = MakeZcylinder([0 0 c]',r,a);        % shape of disk
[vertex2,face2] = MakeZcylinder([0 0 c]',0.01,2*c);   % shape of shaft
uLINK(j).vertex = [vertex1 vertex2];
face2 = face2 + size(vertex1,2);
uLINK(j).face   = [face1 face2];

density = 2.7E+3;                      % density of aluminum [kg/m^2]
uLINK(j).m = pi*r^2*a*density;         % mass of the disk [kg]
uLINK(j).I = [ (a^2 + 3*r^2)/12 0 0;
              0 (a^2 + 3*r^2)/12  0;
              0 0 r^2/2] * uLINK.m;    % inertia tensor of the disk
uLINK(j).c = [0 0 c]';                 % center of mass
```

Fig. 6.27 `MakeTop.m`: Setup a top and its parameters

```
function [vert,face] = MakeZcylinder(pos, radius,len)

a = 10;    % regular polygon for circle
theta = (0:a-1)/a * 2*pi;

x  = radius*cos(theta);
y  = radius*sin(theta);
z1 = len/2 * ones(1,a);
z2 = -z1;

vert    = [x x 0 0;
           y y 0 0;
           z1 z2 len/2 -len/2];     % vertices
for n = 1:3
   vert(n,:) = vert(n,:) + pos(n);
end

face_side = [1:a; a+1:2*a; a+2:2*a a+1; 2:a 1];
face_up   = [1:a; 2:a 1];
face_up(3:4,:) = 2*a+1;  % index of up center
face_down = [a+2:2*a a+1; a+1:2*a];
face_down(3:4,:) = 2*a+2; % index of down center
face = [face_side face_up face_down];
```

Fig. 6.28 MakeZcylinder.m: Create cylinder shape

```
function ret = InvDyn(j)
global uLINK
uLINK(1).dvo = [0 0 0]';
uLINK(1).dw  = [0 0 0]';
if j >= 1 & j <= 3
    uLINK(1).dvo(j) = 1;
elseif j >= 4 & j <= 6
    uLINK(1).dw(j-3) = 1;
end
for n=1:length(uLINK)-1
    if n == j-6
        uLINK(n+1).ddq = 1;
    else
        uLINK(n+1).ddq = 0;
    end
end
ForwardAllKinematics(1);
[f,tau] = InverseDynamics(1);
ret = [f',tau',uLINK(2:end).u]';
```

Fig. 6.29 InvDyn.m: Inverse dynamics by unit vector method

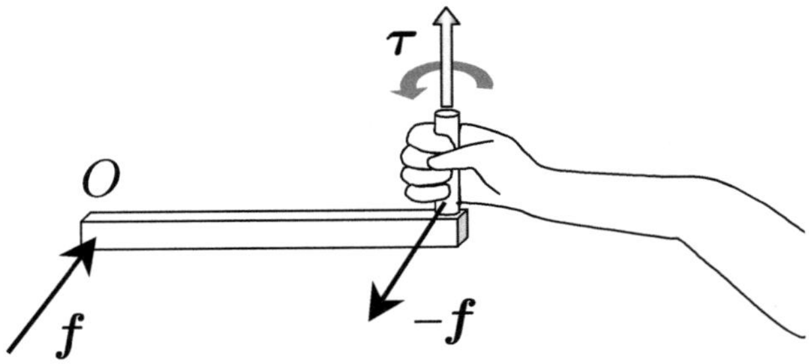

Fig. 6.30 The answer for the quiz of page 207: To keep the rod stationary, a person must apply the force $-\boldsymbol{f}$ as well as the moment $\boldsymbol{\tau}$. If the person who is gripping the handle is on a small boat, he/she will start moving in the direction of vector $\boldsymbol{f}$.

References

1. Aldebaran robotics, `http://www.aldebaran-robotics.com`
2. Carnegie mellon university TARTAN RESCUE, `http://www.rec.ri.cmu.edu/projects/tartanrescue/`
3. DARPA Robotics Challenge, `http://www.theroboticschallenge.org`
4. Products of Kondo Kagaku Co. Ltd. (in Japanese), `http://kondo-robot.com/product/`
5. Broad Agency Announcement DARPA Robotics Challenge, DARPA-BAA-12-39 (April 2012)
6. Takanishi, A., Ishida, M., Yamazaki, Y., Kato, I.: The realization of dynamic walking by the biped walking robot. In: Proceedings of IEEE Int. Conf. on Robotics and Automation, pp. 459–466 (1985)
7. Goswami, A., Espiau, B., Keramane, A.: Limit cycles in a passive compass gait biped and passivity-mimicking control laws. Journal of Autonomous Robots 4(3), 273–286 (1997)
8. Harashima, A.: Mechanics I, II. Shokabo (1972) (in Japanese)
9. Herdt, A., Diedam, H., Wieber, P.-B., Dimitrov, D., Mombaur, K., Diehl, M.: Online walking motion generation with automatic footstep placement. Advanced Robotics 24, 719–737 (2010)
10. Takanishi, A., Ishida, M., Yamazaki, Y., Kato, I.: The realization of dynamic walking by the biped walking robot WL-10RD. In: Proceedings of 1985 International Conference on Advanced Robotics (ICAR), pp. 459–466 (1985)
11. Takanishi, A., Egusa, Y., Tochizawa, M., Takeya, T., Kato, I.: Realization of dynamic biped walking stabilized with trunk motion. In: Proceedings of RoManSy 7: 7th CISM-IFToMM Symposium on Theory and Practice of Robots and Manipulators, pp. 68–79 (1990)
12. Cho, B.K., Park, S.S., Oh, J.H.: Controllers for running in the humanoid robot, HUBO. In: Proceedings of IEEE-RAS International Conference on Humanoid Robots, pp. 385–390 (2009)
13. Lim, B., Lee, J., Kim, J., Lee, M., Kwak, H., Kwon, S., Lee, H., Kwon, W., Roh, K.: Optimal gait primitives for dynamic bipedal locomotion. In: Proceedings of the IEEE/RSJ International Conference on Intelligent Robots and Systems, pp. 4013–4018 (2012)
14. Thuilot, B., Goswami, A., Espiau, B.: Bifurcation and chaos in a simple passive bipedal gait. In: Proceedings of the 1997 IEEE International Conference on Robotics & Automation, pp. 792–798 (1997)

15. DLR (German Aerospace Center). The robot is complete – arms and hands for TORO, the walking machine,
`http://www.dlr.de/dlr/en/desktopdefault.aspx/tabid-10080/150_read-6601/year-2013/150_page-5/#gallery/9208`
16. Ott, C., Baumgärtner, C., Mayr, J., Fuchs, M., Burger, R., Lee, D., Eiberger, O., Albu-Schäffer, A., Grebenstein, M., Hirzinger, G.: Development of a biped robot with torque controlled joints. In: Proceedings of IEEE-RAS International Conference on Humanoid Robots (Humanoids 2010), pp. 167–173 (2010)
17. TOYOTA MOTOR CORPORATION. Partner robot,
`http://www.toyota-global.com/innovation/partner_robot/`
18. Craig, J.J.: Introduction to Robotics: Mechanics and Control, 2nd edn. Addison-Wesley Publishing Company, Inc. (1989)
19. Witt, D.C.: A feasibility study on powerd lower-limb prostheses. In: Proceedings of Symposium on the Basic Problems of Prehension, Movement and Control of Artificial Limbs, pp. 1–8 (1968)
20. de Garis, H.: Gennets: Genetically programmed neural net. In: Proceedings of IJCNN 1991 Singapore. In: Int. Joint Conf. on Neural Networks, pp. 1391–1396 (1991)
21. Hobbelen, D.G.E., Wisse, M.: Active lateral placement for 3D stabilization of a limit cycle walker prototype. International Journal of Humanoid Robotics 6(1), 93–116 (2009)
22. Lahr, D., Hong, D.: The development of CHARLI: A linear actuated powered full size humanoid robot. In: Proceedings of the International Conference on Ubiquitous Robots and Ambient Intelligence (URAI 2008) (November 2008)
23. Nakano, E., Komoriya, K., Yoneda, K., Takahashi, T.: Advanced Mobile Robotics. Kodansha Scientific (2004) (in Japanese)
24. Westervelt, E.R., Grizzle, J.W., Chevallereau, C., Choi, J.H., Morris, B.: Feedback Control of Dynamic Bipedal Robot Locomoation. CRC Press (2007)
25. Neo, E.S., Yokoi, K., Kajita, S., Kanehiro, F., Tanie, K.: A switching command-based whole-body operation method for humanoid robots. IEEE/ASME Tras. Mechatronics 10(5), 2569–2574 (2005)
26. Neo, E.S., Yokoi, K., Kajita, S., Tanie, K.: A framework for remote execution of whole body motions for humanoid robots. In: Proceedings of IEEE-RAS Int. Conf. Humanoid Robots, pp. 58–68 (2004)
27. Miura, H., et al.: Collected articles of biped robots (Nisoku-hokou robot siryou-shuu), 2nd edn. Report of Grants-in-Aid for Scientific Research. Ministry of Education
28. Kanehiro, F., Kaneko, K., Fujiwara, K., Harada, K., Kajita, S., Yokoi, K., Hirukawa, H., Akachi, K., Isozumi, T.: The first humanoid robot that has the same size as a human and that can lie down and get up. In: Proceedings of IEEE Int. Conf. on Robotics and Automation, pp. 1633–1639 (2003)
29. Asano, F., Yamakita, M.: Virtual gravity and coupling control for robotic gait synthesis. IEEE Transactions on Systems, Man and Cybernetics, Part A: Systems and Humans 31(6), 737–745 (2001)
30. Roy Featherstone. Robot Dynamics Algorithms. Kluwer Academic Publishers (1987)

31. Kanehiro, F., Hirukawa, H., Kajita, S.: OpenHRP: Open Architecture Humanoid Robotics Platform. The International Journal o Robotics Research 23(2), 155–165 (2004)
32. Miyazaki, F., Arimoto, S.: A control theoretic study on dynamical biped locomotion. Journal of Dynamic Systems, Measurement, and Control 102, 233–239 (1980)
33. Fujisoft, Inc. PALRO (in japanese), http://palro.jp/
34. Metta, G., Sandini, G., Vernon, D., Natale, L., Nori, F.: The iCub humanoid robot: an open platform for research in embodied cognition. In: Proceedings of the 8th Workshop on Performance Metrics for Intelligent Systems (PerMIS 2008), pp. 50–56 (2008)
35. Nelson, G., Saunders, A., Neville, N., Swilling, B., Bondaryk, J., Billings, D., Lee, C., Playter, R., Raibert, M.: PETMAN: A humanoid robot for testing chemical protective clothing. Journal of the Robotics Society of Japan 30(4), 372–377 (2012)
36. Goldstein, H.: Classical Mechanics, 2nd edn. Addison-Wesley (1980)
37. Goswami, A.: Posturel Stability of Biped Robots and the Foot-Rotation Indicator(FRI) Point. Int. J. of Robotics Research 18(6), 523–533 (1999)
38. Hirukawa, H., Kanehiro, F., Kaneko, K., Kajita, S., Morisawa, M.: Dinosaur robotics for entertainment applications. IEEE Robotics & Automation Magazine 14(3), 43–51 (2007)
39. Hirukawa, H., Hattori, S., Harada, K., Kajita, S., Kanehiro, F., Fujiwara, K., Morisawa, M.: A Universal Stability Criterion of the Foot Contact of Legged Robots - Adios ZMP. In: Proceedings of IEEE International Conference on Robotics and Automation (ICRA 2006), pp. 1976–1983 (2006)
40. Kondo, H., Shimizu, J., Hashimoto, K., Hattori, K., Nishikawa, K., Takezaki, Y., Hama, Y., Yoshimura, Y., Lim, H., Takanishi, A.: Realization of Walking by FFT-based Online Pattern Generation. In: Proceedings of the 12th International Conference on Climbing and Walking Robots and the Support Technologies for Mobile Machines (CLAWAR 2009), pp. 615–622 (September 2009)
41. Miura, H., Shimoyama, I.: Dynamic walk of a biped. International Journal of Robotics Research 3(2), 60–74 (1984)
42. Hyon, S.-H.: Compliant terrain adaptation for biped humanoids without measureing ground suface and contact forces. IEEE Transactions on Robotics 25(1), 171–178 (2009)
43. Honda Motor Co. Inc. NEWS: Honda unveils all-new ASIMO with significant advancements, http://asimo.honda.com/news/honda-unveils-all-new-asimo-with-significant-advancements/newsarticle_0125/
44. Englsberger, J., Ott, C., Roa, M.A., Albu-Schäffer, A., Hirzinger, G.: Bipedal walking control based on capture point dynamics. In: Proceedings of IEEE/RSJ International Conference on Intelligent Robots and Systems (IROS 2011) (2011)
45. Jet Propulsion Laboratory (Caltech). DARPA robotics challenge, RoboSimian (track A), http://www-robotics.jpl.nasa.gov/tasks/showTask.cfm?TaskID=236&tdaID=700043

46. Furusho, J., Sano, A.: Sensor-based control of a nine-link biped. International Journal of Robotics Research 9(2), 83–98 (1990)
47. Furusho, J., Masubuchi, M.: A theoretically motivated reduced order model for the control of dynamic biped locomotion. Journal of Dynamic Systems, Measurement, and Control 109, 155–163 (1987)
48. Kuffner, J.J., LaValle, S.M.: RRT-connect: An efficient approach to single-query path planning. In: Proceedings of IEEE Int. Conf. on Robotics & Automation, pp. 995–1001 (2000)
49. Kuffner, J., Nishiwaki, K., Kagami, S., Inaba, M., Inoue, H.: Motion planning for humanoid robots under obstacle and dynamic balance constraints. In: Proceedings of IEEE Int. Conf. on Robotics & Automation, pp. 692–698 (2001)
50. Pratt, J., Carff, J., Drakunov, S., Goswami, A.: Capture point: A step toward humanoid push recovery. In: Proceedings of IEEE-RAS International Conference on Humanoid Robots (Humanoids 2006), pp. 200–207 (2006)
51. Urata, J., Nishiwaki, K., Okada, K., Kagami, S., Inaba, M.: Online decision of foot placement using singular LQ preview regulation. In: Proceedings of IEEE-RAS International Conference on Humanoid Robots (Humanoids 2011), pp. 13–18 (2011)
52. Urata, J., Nishiwaki, K., Nakanishi, Y., Okada, K., Kagami, S., Inaba, M.: Online walking pattern generation for push recovery and minimum delay to commanded changed of direction and speed. In: Proceedings of IEEE/RSJ International Conference on Intelligent Robots and Systems, pp. 3411–3416 (2012)
53. Yamaguchi, J., Soga, E., Inoue, S., Takanishi, A.: Development of a bipedal humanoid robot - control method of whole body cooperative dynamic biped walking. In: Proceedings of IEEE Int. Conf. on Robotics and Automation, pp. 368–374 (1999)
54. Aida, K., Kitamori, T.: A relation between optimal preview control and least-squares smoothing and a scheme of smoothed inverse system. Transactions of SICE 25(4), 419–426 (1989) (in Japanese)
55. Kailath, T.: Linear Systems. Prentice-Hall, Inc. (1980)
56. Kawada Industries, Inc. Humanoid robot HRP-4, `http://global.kawada.jp/mechatronics/hrp4.html`
57. Doya, K.: Walking pattern learning robot (hokou pattern gakushu robot). Journal of the Robotics Society of Japan 8(3), 117 (1990) (in Japanese)
58. Fujiwara, K., Kanehiro, F., Kajita, S., Yokoi, K., Saito, H., Harada, K., Kaneko, K., Hirukawa, H.: The first human-size humanoid that can fall over safely and stand-up again. In: Proceedings of IEEE/RSJ Int. Conf. on Intelligent Robots and Systems, pp. 1920–1926 (2003)
59. Hara, K., Yokogawa, R., Sadao, K.: Dynamic control of biped locomotin robot for disturbance on lateral plane. In: Proceedings of the Japan Society of Machanical Engineers 72nd Kansai Meeting, vol. 38, pp. 10–37 (1998) (in Japanese)
60. Harada, K., Kajita, S., Kaneko, K., Hirukawa, H.: An analytical method for real-time gait planning for humanoid robots. International Journal of Humanoid Robotics 3(1), 1–19 (2006)

61. Hase, K., Yamazaki, N.: Computer simulation study of human locomotion with a three-dimensional entire-body neuro-musculo-skeletal model. JSME International Journal, Series C 45(4), 1040–1072 (2002)
62. Hirai, K., Hirose, M., Haikawa, Y., Takenaka, T.: The development of honda humanoid robot. In: Proceedings of the 1998 IEEE International Conference on Robotics & Automation, pp. 1321–1326 (1998)
63. Kaneko, K., Kanehiro, F., Morisawa, M., Miura, K., Nakaoka, S., Kajita, S.: Cybernetic human HRP-4C. In: Proceedings of 9th IEEE-RAS International Conference on Humanoid Robots, pp. 7–14 (2009)
64. Kaneko, K., Kanehiro, F., Morisawa, M., Tsuji, T., Miura, K., Nakaoka, S., Kajita, S., Yokoi, K.: Hardware improvement of cybernetic human HRP-4C towards entertainment use. In: Proceedings of IEEE/RSJ International Conference on Intelligent Robots and Systems, pp. 4392–4399 (2011)
65. Kaneko, K., Kanehiro, F., Kajita, S., Hirukawa, H., Kawasaki, T., Hirata, M., Akachi, K., Isozumi, T.: Humanoid robot HRP-2. In: Proceedings of IEEE International Conference on Robotics and Automation, pp. 1083–1090 (2004)
66. Kaneko, K., Kanehiro, F., Kajita, S., Yokoyama, K., Akachi, K., Kawasaki, T., Ota, S., Isozumi, T.: Design of prototype humanoid robotics platform for HRP
67. Kaneko, K., Harada, K., Miyamori, G., Akachi, K.: Humanoid robot HRP-3. In: Proceedings of 2008 IEEE/RSJ International Conference on Intelligent Robots and Systems, pp. 2471–2478 (2008)
68. Koganezawa, K., Takanisi, A., Sugano, S. (eds.): Development of Waseda robot — The study of biomechanisms at Kato Laboratory, 3rd edn. Ichiro Kato Laboratory (1991) (in Japanese)
69. Miura, K., Morisawa, M., Kanehiro, F., Kajita, S., Kaneko, K., Yokoi, K.: Proceedings of IEEE/RSJ International Conference on Intelligent Robots and Systems, pp. 4428–4435 (2011)
70. Nagasaka, K., Inaba, M., Inoue, H.: Stabilization of dynamic walk on a humanoid using torso position compliance control. In: Proceedings of 17th Annual Conference of the Robotics Society of Japan, pp. 1193–1194 (1999) (in Japenese)
71. Narioka, K., Tsugawa, S., Hosoda, K.: 3D limit cycle walking of musculoskeletal humanoid robot with flat feet. In: Proceedings of the IEEE/RSJ International Confernce on Intelligent Robots and Systems, pp. 4676–4681 (2009)
72. Nishiwaki, K., Kagami, S.: Sensor feedback modification methods that are suitable for the short cycle pattern generation of humanoid walking. In: Proceedings of IEEE/RSJ International Conference on Intelligent Robots and Systems (IROS 2007), pp. 4214–4220 (2007)
73. Nishiwaki, K., Kagami, S., Kuniyoshi, Y., Inaba, M., Inoue, H.: Online generation of desired walking motion on humanoid based on a fast method of motion pattern that follows desired ZMP. In: Proceedings of 19th Annual Conference of Robotics Society of Japan, pp. 985–986 (2001) (in Japanese)
74. Osuka, K., Kirihara, K.: Motion analysis and experiments of passive walking robot Quartet II. In: Proceedings of IEEE International Conferene on Robotics & Automation, pp. 3052–3056 (2000)
75. Sorao, K., Murakami, T., Ohnishi, K.: Walking control of a biped robot by impedance control. Transaction of IEE of Japan 117-D(10), 1227–1233 (1997) (in Japanese)

76. Yamane, K.: Simulating and Generating Motions of Human Figures. Springer (2004)
77. Yamane, K., Nakamura, Y.: Dynamics computation of structure-varying kinematic chains for motion synthesis of humanoid. In: Proceedings of IEEE International Conference on Robotics and Automation (ICRA1999), pp. 714–721 (1999)
78. Yamane, K., Nakamura, Y.: Efficient parallel dynamics computation of human figures. In: Proceedings of the 2002 IEEE International Conference on Robotics & Automation, pp. 530–537 (2002)
79. Yamane, K., Nakamura, Y.: Dynamics filter–concept and implementation of online motion generator for human figures. IEEE Trans. on Robotics & Automation 19(3), 421–432 (2003)
80. Yoneda, K., Hirose, S.: Tumble stability criterion of integrated locomotion and manipulation. In: Proceedings of IEEE/RSJ International Conference on Intelligent Robots and Systems 1996 (IROS 1996), pp. 870–876 (1996)
81. Vukobratović, M., Borovac, B., Surla, D., Stokić, D.: Biped Locomotion — Dynamics, Stability, Control and Application. Springer (1990)
82. Hayase, M., Ichikawa, K.: Optimal servosystem utilizing future value of desired function. Transactions of SICE 5(1), 86–94 (1969) (in Japanese)
83. Hirose, M., Takenaka, T., Gomi, H., Ozawa, N.: Humanoid robot. Journal of the Robotics Society of Japan 15(7), 983–985 (1997) (in Japenese)
84. Kumagai, M., Tomita, H., Emura, T.: Sensor-based walking of human type biped robot – 2nd report, active control of body attitude. In: Proceedings of JSME Conference on Robotics and Mechatronics (ROBOMEC 1998), pp. 2CIII1–2 (1998) (in Japanese)
85. Morisawa, M., Kanehiro, F., Kaneko, K., Mansard, N., Sola, J., Yoshida, E., Yokoi, K., Laumond, J.-P.: Combining suppression of the disturbance and reactive stepping for recovering balance. In: Proceedings of IEEE/RSJ International Conference on Intelligent Robots and Systems, pp. 3150–3156 (2010)
86. Tomizuka, M., Rosenthal, D.E.: On the optimal digital state vector feedback controller with integral and preview actions. Transaction of the ASME, Journal of Dynamic Systems, Measurement, and Control 101, 172–178 (1979)
87. Uchiyama, M., Nakamura, Y.: Iwanami Lecture of Robotics 2: Robot Motion. Iwanami Shoten (2004) (in Japanse)
88. Vukobratović, M., Borovac, B.: Zero-Moment Point — Thirty Five Years on its Life. International Journal of Humanoid Robotics 1(1), 157–173 (2004)
89. Vukobratović, M., Borovac, B., Šurdilović, D.: Zero-moment point — proper interpreation and new applications. In: Proceedings of IEEE-RAS Intr. Conf. on Humanoid Robots, pp. 237–244 (2001)
90. Vukobratović, M., Stepanenko, J.: On the stability of anthropomorphic systems. Mathematical Biosciences 15, 1–37 (1972)
91. Spong, M.W., Bullo, F.: Controlled symmetries and passive walking. IEEE Transactions on Automatic Control 50(7), 1082–1085 (2005)
92. Walker, M.W., Orin, D.E.: Efficient dynamics computer simulation of robotic mechanisms. Journal of Dynamic Systems, Measurement, and Control 104, 205–211 (1982)

93. Nagasaka, K.: Whole body motion generation for humanoid robot by dynamics filter (Dourikigaku-filter niyoru ningengata-robot no zenshin-unndou seisei). The University of Tokyo (2000) (in Japanese)
94. Napoleon, N., Izu, H., Nakaura, S., Sampei, M.: An analysis of ZMP control problem of humanoid robot with compliances in sole of the foot. In: Proceedings of the 16th IFAC World Congress (2005)
95. Sugimoto, N., Morimoto, J., Hyon, S.-H., Kawato, M.: The eMOSAIC model for humanoid robot control. Neural Networks 30(0), 8–19 (2012)
96. Paul, R.P.: Robot Manipulators: Mathematics, Programming, and Control. MIT Press (1981)
97. Raibert, M.H.: Legged robots that balance. MIT Press, Cambridge (1986)
98. Featherstone, R.: A divide-and-conquer articulated-body algorithm for parallel $O(log(n))$ calculation of rigid-body dynamics. International Journal of Robotics Research 18(9), 867–892 (1999)
99. Featherstone, R.: The acceleration vector of a rigid body. The International Journal of Robotics Research 20(11), 841–846 (2001)
100. Katoh, R., Mori, M.: Control method of biped locomotion giving asymtotic stability of trajectory. Automatica 20(4), 405–414 (1984)
101. Murray, R.M., Li, Z., Sastry, S.S.: A Mathematical Introduction to Robotics Manipulation. CRC Press (1994)
102. PAL Robotics. REEM-C, `http://www.pal-robotics.com/robots/reem-c`
103. ROBOTIS. Open platform humanoid project, `http://www.robotis.com/xe/darwin_en`
104. Rockafellar, R.T.: Convex Analysis. Princeton University Press (1970)
105. Feynman, R.P., Leighton, R.B., Sands, M.L.: ch. 52. Addison-Wesley (1965)
106. Tedrake, R., Zhang, T.W., Seung, H.S.: Stochastic policy gradient reinforcement learning on a simple 3D biped. In: Proceedings of 2004 IEEE/RSJ International Conference on Intelligent Robots and Systems (IROS 2004), pp. 2849–2854 (2004)
107. Hyon, S.-H., Morimoto, J., Kawato, M.: From compliant balancing to dynamic walking on humanoid robot: Integration of CNS and CPG. In: Proceedings of 2010 IEEE International Conference on Robotics and Automation, pp. 1084–1085 (2010)
108. Kajita, S., Kanehiro, F., Fujiwara, K., Harada, K., Yokoi, K., Hirukawa, H.: Resolved momentum control: Humanoid motion planning based on the linear and angular momentum. In: Proceedings of IEEE/RSJ Int. Conf. on Intelligent Robots and Systems, pp. 1644–1650 (2003)
109. Arimoto, S., Miyazaki, F.: A hierachical control scheme for biped robots. Journal of the Robotics Society of Japan 1(3), 167–175 (1983) (in Japanese)
110. Collins, S., Ruina, A., Tedrake, R., Wisse, M.: Efficient bipedal robots based on passive-dynamic walkers. Science 307, 1082–1085 (2005)
111. Collins, S.H., Wisse, M., Ruina, A.: A three-dimensional passive-dynamic walking robot with two legs and knees. International Journal of Robotics Research 20(7), 607–615 (2001)
112. Kagami, S., Kanehiro, F., Tajima, Y., Inaba, M., Inoue, H.: Autobalancer: An online dynamic balance compensation scheme for humanoid robot. In: Proceedings of 4th Int. Workshop on Algorithmic Foundations on Robotics, pp. SA79–SA89 (2000)

113. Kagami, S., Nishiwaki, K., Kuffner Jr., J.J., Kuniyoshi, Y., Inaba, M., Inoue, H.: Design and implementation of software research platform for humanoid robotics: H7. In: Proceedings of the 2001 IEEE-RAS International Conference on Humanoid Robots, pp. 253–258 (2001)
114. Kagami, S., Nishiwaki, K., Kitagawa, T., Sugihara, T., Inaba, M., Inoue, H.: A fast generation method of a dynamically stable humanoid robot trajectory with enhanced ZMP constraint. In: Proceedings of IEEE International Conference on Humanoid Robot (2000)
115. Kajita, S., Tani, K.: Study of dynamic walk control of a biped robot on rugged terrain — derivation and application of the linear inverted pendulum mode. Journal of Robotics and Mechatronics 5(6), 516–523 (1993)
116. Kajita, S., Tani, K.: Experimental study of biped dynamic walking. IEEE Control Systems 16(1), 13–19 (1996)
117. Kajita, S., Morisawa, M., Miura, K., Nakaoka, S., Harada, K., Kaneko, K., Kanehiro, F., Yokoi, K.: Biped walking stabilization based on linear inverted pendulum tracking. In: Proceedings of IEEE/RSJ International Conference on Intelligent Robots and Systems (IROS 2010), pp. 4489–4496 (2010)
118. Kajita, S., Matsumoto, O., Saigo, M.: Real-time 3D walking pattern generation for a biped robot with telescopic legs. In: Proceedings of the IEEE International Conference on Robotics and Automation, pp. 2299–2306 (2001)
119. Lohmeier, S., Buschmann, T., Ulbrich, H.: Humanoid robot LOLA. In: Proceedings of IEEE International Conference on Robotics and Automation, pp. 775–780 (2009)
120. Masaki, S., Shimizu, M., Endo, K., Furuta, T., Okumura, Y., Tawaara, T., Kitano, H.: A foot sensor using multiple independent force sensors for compact-sized humanoid robots. In: Proceedings of ROBOMEC (2003) (in Japanese)
121. Nakaoka, S., Nakazawa, A., Yokoi, K., Hirukawa, H., Ikeuchi, K.: Generating whole body motions for a biped humanoid robot from captured human dances. In: Proceedings of IEEE Int.Conf. on Robotics & Automation, pp. 3905–3910 (2003)
122. Nakaoka, S., Kajita, S., Yokoi, K.: Intuitive and flexible user interface for creating whole body motions of biped humanoid robots. In: Proceedings of IEEE/RSJ International Conference on Intelligent Robots and Systems, pp. 1675–1682 (2010)
123. Nozaawa, S., Ueda, R., Kakiuchi, Y., Okada, K., Inaba, M.: A full-body motion control method for a humanoid robot based on on-line estimation of the operational force of an object with an unknown weight. In: Proceedings of IEEE/RSJ International Conference on Intelligent Robots and Systems, pp. 2684–2691 (2010)
124. Sugihara, T.: Solvability-unconcerned inverse kinematics based on Levenberg-Marquardt method with robust damping. In: Proceedings of 9th IEEE-RAS International Conference on Humanoid Robots, pp. 555–560 (2009)
125. Sugihara, T.: Standing Stabilizability and Stepping Maneuver in Planar Bipedalism based on the Best COM-ZMP Regulator. In: IEEE International Conference on Robotics and Automation (ICRA 2009) (2009)
126. Taga, G., Yamaguchi, Y., Shimizu, H.: Self-organized control of bipedal locomotion by neural oscillators in unpredictable environment. Biological Cybernetics (65), 147–159 (1991)
127. Takano, M.: Robot Kinematics. Ohmsha (2004) (in Japanese)

128. Sheridan, T.B.: Three models of preview control. IEEE Transaction on Human Factors in Electronics 7(2), 91–108 (1966)
129. Furuta, T., Tawara, T., Okumura, Y., Shimizu, M., Shimomura, M., Endo, K., Yamanaka, S., Kitano, H.: Morph3: A compact-size humanoid robot system with acrobatic behavior capability. In: Proceedings of ROBOMEC (2003) (in Japanese)
130. Katayama, T., Ohki, T., Inoue, T., Kato, T.: Desing of an optimal controller for a discrete time system subject to previewable demand. International Journal of Control 41(3), 677–699 (1985)
131. McGeer, T.: Passive dynamic walking. The International Journal of Robotics Research 9(2), 62–82 (1990)
132. McGeer, T.: Passive walking with knees. In: Proceedings of IEEE Int. Conf. on Robotics & Automation, vol. 3, pp. 1640–1645 (1990)
133. Mita, T., Yamaguchi, T., Kashiwase, T., Kawase, T.: Realization of a high speed biped using modern control theory. International Journal of Control 40(1), 107–119 (1984)
134. Sugihara, T., Nakamura, Y., Inoue, H.: Realtime humanoid motion generation through ZMP manipulation based on inverted pendulum control. In: Proceedings of the IEEE International Conference on Robotics and Automation, pp. 1404–1409 (2002)
135. Wampler, C.W.: Manipulator inverse kinematic solutions based on vector formulations and damped least-square methods. IEEE Transactions on Systems, Man, and Cybernetics 16(1), 93–101 (1986)
136. Press, W.H., Flannery, B.P., Teukolsky, S.A., Vetterling, W.T.: Numerical Recipes in C. Cambridge University Press (1988)
137. Wieber, P.-B.: Trajectory Free Linear Model Predictive Control for Stable Walking in the Presence of Strong Perturbations, pp. 137–142 (2006)
138. Yakamura, Y., Hanafusa, H.: Inverse kinematic solutions with singularity robustness for robot manipulator control. Journal of Dynamic Systems, Measurement and Control 108, 163–171 (1986)
139. Choi, Y., Kim, D., You, B.-J.: On the walking control for humanoid robot based on the kinematic resolution of COM Jacobian with embedded motion. In: Proceedings of the 2006 IEEE International Conference on Robotics and Automation (ICRA 2006), pp. 2655–2660 (2006)
140. Fujimoto, Y., Kawamura, A.: Three dimensional digital simulation and autonomous walking control for eight-axis biped robot. In: Proceedings of IEEE International Conference on Robotics and Automation (ICRA 1995), pp. 2877–2884 (1995)
141. Ikemata, Y., Sano, A., Fujimoto, H.: A physical principle of gait generation and its stabilization derived from mechanism of fixed point. In: Proceedings of IEEE International Conference on Robotics & Automation, pp. 836–841 (2006)
142. Ogura, Y., Shimomura, K., Kondo, H., Morishima, A., Okubo, T., Momoki, S., Lim, H., Takanishi, A.: Human-like walking with knee stretched, heel-contact and toe-off motion by a humanoid robot. In: Proceedings of IEEE/RSJ International Conference on Intelligent Robots and Systems, pp. 3976–3981 (2006)

143. Okumura, Y., Tawara, T., Endo, K., Furuta, T., Shimzu, M.: Realtime ZMP compensation for biped walking robot using adaptive inertia force control. In: Proceedings of the 2003 IEEE/RSJ International Conference on Intelligent Robots and Systems (IROS 2003), pp. 335–339 (2003)
144. Yoshikawa, T.: Foundations of Robotics: Analysis and Control. MIT Press (1990)
145. Yoshino, R.: Stabilizing control of high-speed walking robot by walking pattern regulator. Journal of the Robotics Society of Japan 18(8), 1122–1132 (2000) (in Japanese)
146. Sumi, Y., Kawai, Y., Yoshimi, T., Tomita, F.: 3D object recognition in cluttered environments by segment-based stereo vision. Int. J. Computer Vision 46(1), 5–23 (2002)
147. Tamiya, Y., Inaba, M., Inoue, H.: Realtime balance compensation for dynamic motion of full-body humanoid standing on one leg. Journal of Robotics Society Japan 17(2), 268–274 (1999) (in Japanese)
148. Peng, Z., Fu, Y., Tang, Z., Huang, Q., Xiao, T.: Online walking pattern generation and system software of humanoid BHR-2. In: Proceedings of IEEE/RSJ International Conference on Intelligent Robot and Systems, pp. 5471–5476 (2006)

Index

Shuuji Kajita · Hirohisa Hirukawa
Kensuke Harada · Kazuhito Yokoi

Introduction to Humanoid Robotics

Springer

Shuuji Kajita
National Institute of Advanced
Industrial Science & Technology (AIST)
Intelligent Systems Research
Institute Humanoid Research Group
Tsukuba Central 2
1-1-1 Umezono, Tsukuba, Ibaraki
305-8568 Japan
E-mail: s.kajita@aist.go.jp

Hirohisa Hirukawa
National Institute of Advanced
Industrial Science & Technology (AIST)
Intelligent Systems Research
Institute Humanoid Research Group
Tsukuba Central 2
1-1-1 Umezono, Tsukuba, Ibaraki
305-8568 Japan
E-mail: hiro.hirukawa@aist.go.jp

Kensuke Harada
National Institute of Advanced
Industrial Science & Technology (AIST)
Intelligent Systems Research
Institute Humanoid Research Group
Tsukuba Central 2
1-1-1 Umezono, Tsukuba, Ibaraki
305-8568 Japan
E-mail: kensuke.harada@aist.go.jp

Kazuhito Yokoi
National Institute of Advanced
Industrial Science & Technology (AIST)
Intelligent Systems Research
Institute Humanoid Research Group
Tsukuba Central 2
1-1-1 Umezono, Tsukuba, Ibaraki
305-8568 Japan
E-mail: kazuhito.yokoi@aist.go.jp

ISSN 1610-7438 ISSN 1610-742X (electronic)
ISBN 978-3-642-54535-1 ISBN 978-3-642-54536-8 (eBook)
DOI 10.1007/978-3-642-54536-8
Springer Heidelberg New York Dordrecht London

Library of Congress Control Number: 2014932049

Translation from the Japanese language edition: *Humanoid Robot* by Shuji Kajita

Printed on acid-free paper

Springer is part of Springer Science+Business Media (www.springer.com)

MIX
Papier aus verantwortungsvollen Quellen
Paper from responsible sources
FSC® C105338

If you have any concerns about our products,
you can contact us on
ProductSafety@springernature.com

In case Publisher is established outside the EU,
the EU authorized representative is:
Springer Nature Customer Service Center GmbH
Europaplatz 3, 69115 Heidelberg, Germany

Printed by Libri Plureos GmbH
in Hamburg, Germany